Probability and the art of judgment

Richard Jeffrey is indisputably one of the most distinguished and influential philosophers working in the field of decision theory and the theory of knowledge. His work is distinctive in showing the interplay of epistemological concerns with probability and utility theory. Not only has he made use of standard probabilistic and decision theoretic tools to clarify concepts of evidential support and informed choice, he has also proposed significant modifications of the standard Bayesian position in order that it provide a better fit with actual human experience. Probability logic is viewed not as a source of judgment but as a framework for explaining the implications of probabilistic judgments and testing their mutual compatibility.

This collection of essays spans a period of some 35 years and includes what have become some of the classic works in the literature. There is also one completely new piece, and in many instances Jeffrey includes afterthoughts on the older essays. The volume will be of particular interest to epistemologists, philosophers of science, probabilists, statisticians, decision theorists, and judgmental psychologists.

Cambridge Studies in Probability, Induction, and Decision Theory

General editor: Brian Skyrms

Advisory editors: Ernest W. Adams, Ken Binmore, Persi Diaconis, William L. Harper, John Harsanyi, Richard C. Jeffrey, Wolfgang Spohn, Patrick Suppes, Amos Tversky, Sandy Zabell

This new series is intended to be the forum for the most innovative and challenging work in the theory of rational decision. It focuses on contemporary developments at the interface between philosophy, psychology, economics, and statistics. The series addresses foundational theoretical issues, often quite technical ones, and therefore assumes a distinctly philosophical character.

Probability
and
the art of judgment

Richard Jeffrey

Department of Philosophy, Princeton University

CAMBRIDGE
UNIVERSITY PRESS

Published by the Press Syndicate of the University of Cambridge
The Pitt Building, Trumpington Street, Cambridge CB2 1RP
40 West 20th Street, New York, NY 10011-4211, USA
10 Stamford Road, Oakleigh, Victoria 3166, Australia

First published 1992

Library of Congress Cataloging-in-Publication Data
Jeffrey, Richard C.
Probability and the art of judgment / Richard Jeffrey.
p.   cm. – (Cambridge studies in probability, induction, and
decision theory)
ISBN 0-521-39459-7. – ISBN 0-521-39770-7 (pbk.)
1. Probabilities.   2. Judgment.   I. Title.
II. Series.
BC141.J44   1992
121'.63 – dc20   91–34257
CIP

A catalog record for this book is available from the British Library.

ISBN 0-521-39459-7 hardback
ISBN 0-521-39770-7 paperback

Transferred to digital printing 2003

For Peter

Contents

statistics. A realistic sample space. Exchangeability.
Markov exchangeability. Partial exchangeability.
Conclusion.   117

Preface

These essays explore a variety of topics, ranging from decision theory and the philosophy of mind to epistemology and scientific methodology, from a probabilistic viewpoint of a sort called subjective (Bruno de Finetti's term), personal (L. J. Savage's), and judgmental (mine). While the focus varies from historical and philosophical overviews to close technical studies, the point of view remains pretty steady (or, anyway, evolves pretty slowly). Its individuating features are a radical probabilism that accepts probability judgments as basic where the usual Bayesian position would root them in certainties, and a view of making up your mind as covering dynamics: how to change your mind. Thus, empirical data may well be coded not as probabilities or as odds but as factors to multiply probabilities or odds by in order to update them in the light of experience. The formal logic of that process identifies invariance of conditional probabilities as validating a certain generalization of the usual "Bayesian" updating scheme, but the judgment of invariance in particular cases is no formal matter. Probability logic is seen not as a source of judgment but as a framework for exploring the implications of probabilistic judgments and testing their mutual compatibility.

Essay 1 is new; the rest, going back over 35 years, appear here essentially unchanged, with afterthoughts clearly labeled, and mostly segregated in postscripts. Here is a brief guide.

1. "Radical probabilism" (1991) depicts the position elaborated here as a development out of logical empiricism, radicalizing Carnap's (and Reichenbach's) probabilism, probabilizing Quine's deductivism.

2. "Valuation and acceptance of scientific hypotheses" (1956) argues that the job of the scientist is not to accept and reject hypotheses but to assign them probabilities.

3. "Probable knowledge" (1968) promotes a generalization of

ix

conditioning ("probability kinematics") as a way of revising probability judgments in the light of uncertain data. Sited in a framework for practical deliberation, this is probabilism of the commonplace, on which probabilistic judgments about large questions impinge only indirectly.

4. "Probability and the art of judgment" (1985) bases an account of scientific method on such impingements, surveys the historical and philosophical context, and stakes some territorial claims.

5. "Bayesianism with a human face" (1983) floats a liberal version of probabilistic methodology. It defends probability kinematics against claims that ordinary conditioning is *the* right way to update probabilities, defends incomplete and indefinite probability judgments as useful probabilistic states of mind, considers how theories can gain probability by explaining previously known facts, and backs de Finetti's view of the probability calculus as the logic of the probable.

6. Addressing puzzles about hard data in the framework of radical probabilism, "Alias Smith and Jones: the testimony of the senses" (1987) offers versions of probability kinematics in which the order of successive probability revisions makes no difference. Those versions code data as factors for revising probabilities or odds, as in essay 1.

7. "Conditioning, kinematics, and exchangeability" (1988) surveys connections between probability kinematics and recent developments in Bayesian statistics extending de Finetti's work on partial exchangeability. Probability kinematics turns out to be equivalent to ordinary conditioning on new probabilities. A solution is offered to Bas van Fraassen's puzzle about the relationship ("the reflection principle") between current and anticipated probabilities. Rigidity conditions, under which it is appropriate to use probability kinematics, turn out to be identical with statistical sufficiency conditions. The study of successive updating begun in essay 6 is taken further.

8. "Preference among preferences" (1974) analyzes weakness of will as real preference for the worse, accompanied by real preference for preferring the better.

9, 10. "On interpersonal utility theory" (1971) and "Remarks on interpersonal utility theory" (1974) explore a solution to the problem of interpersonal comparisons. The second of these essays, written to illuminate the first, might be read first.

11. The 1960s saw the emergence of convincing accounts of randomness as algorithmic incompressibility in the finite case, and of other convincing accounts for infinite sequences. "Mises redux" (1977) rebuts the view that these finally make sense of von Mises' notion of an irregular collective, and thus establish the credentials of a frequency view of probability.

12. "Statistical explanation vs. statistical relevance" (1969) offers an alternative to Hempel's view of probabilistic explanation as inductive argument.

13, 14, 15, 16. "New foundations for Bayesian decision theory" (1965) summarizes the account of preference set forth in the first seven chapters of my book *The Logic of Decision* (1965, 1983), and "Frameworks for preference" (1974) contrasts it favorably with some of its competitors. "Axiomatizing the logic of decision" (1978) and "A note on the kinematics of preference" (1977) are close looks at some of its features.

Fair parts of the work represented here were supported by the History and Philosophy of Science section of the National Science Foundation, the John Simon Guggenheim Foundation, and the National Endowment for the Humanities.

This collection is gratefully dedicated to my dear friend and teacher C. G. (Peter) Hempel.

1

Introduction: Radical probabilism

Adopting a central feature of Stoic epistemology, Descartes treated belief as action that might be undertaken wisely or rashly, and enunciated a method for avoiding false belief, a discipline of the will "to include nothing more in my judgments than what presented itself to my mind with such clarity and distinctness that I would have no occasion to put it in doubt."[1] He called such acts of the will "affirmations," i.e., acts of accepting sentences or propositions as true. (Essay 2 argues against "cognitive" uses of decision theory to choose among such replacements of considered probabilities by specious certainties.)

What do "belief," "acceptance," and "affirmation" mean in this context? I don't know. I'm inclined to doubt that anyone else does, either, and to explain the general unconcern about this lack of understanding by familiarity cf the acceptance metaphor masquerading as intelligibility, perhaps as follows: "Since it's clear enough what's meant by accepting other things – gifts, advice, apologies – and it's clear enough what's meant by sentences' being true, isn't it clear what's meant by accepting sentences as true? Doesn't Quine make 'holding' sentences true the very pivot of his epistemology? And isn't affirmation just a matter of saying 'Yes'?"

The ("Bayesian") framework explored in these essays replaces the two Cartesian options, affirmation and denial, by a continuum of judgmental probabilities in the interval from 0 to 1, endpoints included, or – what comes to the same thing – a continuum of judgmental odds in the interval from 0 to ∞, endpoints included. Zero and 1 are probabilities no less than $1/2$ and $99/100$ are. Probability 1 corresponds to infinite odds, $1:0$. That's a reason for thinking in terms of odds: to remember how momentous it may be to assign probability 1 to a hypothesis. It means you'd stake your all on its

---

1. *Discourse on the method* . . . , part 2.

1

truth, if it's the sort of hypothesis you can stake things on. To assign 100% probability to success of an undertaking is to think it advantageous to stake your life upon it in exchange for any petty benefit. We forget that when we imagine that we'd assign probability 1 to whatever we'd simply state as true.[2]

What is involved in attributing particular judgmental probabilities to sentences? Essays 13 and 14 answer in terms of a theory of preference seen as a relation between sentences or propositions: preference for truth of one sentence ("Cameroon wins") to truth of another ("Britain wins"). This theory is subjectivistic in addressing only the effects of such probability judgments, without saying how those judgments ought to be arrived at. The theory doesn't prejudge attempts like Carnap's to supply norms for forming such judgments; and indeed Carnap accepted this subjectivistic theory as an account of how judgmental probabilities are to be applied, once formed.[3]

Broadly speaking, a Bayesian is a probabilist, a person who sees making up the mind as a matter of either adopting an assignment of judgmental probabilities or adopting certain features of such an assignment, e.g., the feature of assigning higher conditional probability to 5-year survival on a diagnosis of ductal cell carcinoma than on a diagnosis of islet cell carcinoma. Some insist on restricting the term "Bayesian" narrowly to those who see conditioning (or "conditionalization") as the only rational way to change the mind; I don't. (Essays 3, 4, 5.) Rationalistic Bayesianism – hereafter, "rationalism" – is a subspecies of the narrow Bayesianism just noted, according to which there exists a (*logical, a priori*) probability distribution that would define the state of mind of a perfect intelligence, innocent of all experience. Notable subscribers: Bayes,

2. As probabilities p range over the unit interval $[0,1]$, the corresponding odds $o = p/(1 - p)$ range over the extended nonnegative reals $[0,\infty]$, enhancing resolution high in the scale. Thus, probabilities 99%, 99.9%, 99.99% correspond to odds 99, 999, 9999. But at the low end, where odds $p/(1 - p)$ are practically equal to probabilities p, there is no such increase in resolution. Logarithms of odds, increasingly positive above $p = .5$ and symmetrically negative below, yield the same resolution at both ends. (Odds of 99, 999, 9999 become log odds of approximately 2, 3, 4, and at the low end, probabilities of .01, .001 .0001 become log odds of approximately -2, -3, -4.)

3. In his introduction to F. P. Ramsey's *Philosophical Papers* (Cambridge University Press, 1990, p. xviii) D. H. Mellor, who ought to know better, says that "modern Bayesian decision theory (e.g., Jeffrey's *The Logic of Decision* 1965), tells us to 'act in the way we think most likely to realize the objects of our desires' whether or not those thoughts and desires are either reasonable or right."

2

Laplace, W. E. Johnson, J. M. Keynes, Carnap, John Harsanyi. Rationalism and empiricism are two sides of the same Bayesian coin. One side is a purely rational, "logical" element, a prior probability assignment **M** characterizing the state of mind of a newborn Laplacean intelligence. Carnap spent his last 25 years trying to specify **M**. The other side is a purely empirical element, a comprehensive report D of all experience to date. Together, these determine the experienced Laplacean intelligence's judgmental probabilities, obtained by conditioning the "ignorance prior" **M** by the *Protokollsatz D*. Thus $M(H|D)$ is the correct probabilistic judgment about H for anyone whose experiential data base is D.

Radical probabilism makes no attempt to analyze judgment into a purely rational component and a purely empirical component, without residue. It rejects the empiricist myth of the sensuously given data proposition D as well as the rationalist myth of the ignorance prior **M**; it rejects the picture of judgment as a coin with empirical obverse and rational reverse. Let's see why.

On the empirical side, reports of conscious experience are too thin an abstract of our sensory inputs to serve adequately as the first term of the equation

(1) $\qquad\qquad Experience\ +\ reason\ =\ judgment.$

Example: Blindsight.[4] In humans, monkeys, etc., some 90% of optic nerve fibers project to the striate cortex at the very back of the brain via the dorsal lateral geniculate nucleus in the midbrain. But while the geniculo-striate pathway constitutes the major portion of the optic nerve . . . there are at least 6 other branches that end up in the midbrain and subcortical regions . . . , and one of these contains about 100 000 fibres, by no means a trivial pathway – it is larger than the whole of the auditory nerve. . . . Therefore, if striate cortex is removed or its direct input blockaded, one should expect that some visual capacity should remain because all of those non-geniculo-striate pathways are left intact. The paradox is that in man this is usually not so: destruction of occipital cortex . . . characteristically causes blindness in that part of the visual field corresponding to the precise projection map of the retina on to the striate cortex. . . . Admittedly, some primitive visual reflexes can be de-

---

4. *Blindsight, a Case Study and Implications*, by L. Weiskrantz, Clarendon Press, Oxford, 1986, pp. 3–6, 24, and 168–169. See also Patricia Smith Churchland, *Neurophilosophy*, MIT Press, Cambridge, Mass., 1986, pp. 224–228.

tected . . . but typically the patient himself does not appear to discriminate visual events or to have any awareness of them.

The non-geniculo-striate 10% of optic nerve fibres seem to provide visual capacities of which such patients are unaware – capacities which they dismiss as mere guesswork, and which the rest of us need never distinguish as a special category. Thus, although a patient ("D. B.") whose right occipital lobe had been surgically removed "could not see one's outstretched hand, he seemed to be able to reach for it accurately. We put movable markers on the wall to the left of his fixation, and again he seemed to be able to point to them, although he said he did not actually see them. Similarly, when a stick was held up in his blind field either in a horizontal or vertical position, and he was asked to guess which of these two orientations it assumed, he seemed to have no difficulty at all, even though again he said he could not actually see the stick." This sort of thing looks like bad news for the New Way of Ideas, empiricism based on sense data: D. B. has factual sensitivity, a basis for probabilistic judgment, with no corresponding phenomenal sensitivity.

If sense data won't do for the first term of formula (1), perhaps the triggering of sensory receptors will. Quine seems to think so: "By the stimulation undergone by a subject on a given occasion I just mean the temporally ordered set of all those of his exteroceptors that are triggered on that occasion."[5] The experiential data base D might then correspond to an ordered set of such ordered sets, whence the Carnapian judgmental probability $M(H|D)$ might be calculable. But no; not even "in principle." The trouble is that the temporal record of exteroception makes no perceptual sense without a correlated record of interoception; thus, interpretation of a record of activity in the optic nerve requires a correlated record of relative orientations of eye, head, trunk, etc. When Quine's bit of neurophysiology is put in context, his exteroceptive data base looks no more adequate for its purpose than did the sense data base it replaced.

Example· Proprioception and visual perception.[6] Exteroceptive nervous activity is interpreted in the light of concurrent interoceptive

5. W. V. Quine, *The Pursuit of Truth*, Harvard University Press, Cambridge, Mass., 1990, p. 2.
6. J. Allen Hobson, *The Dreaming Brain*, Basic Books, New York, 1988, pp. 110–112.

activity. Thus, optic nerve input is interpreted in the light of concurrent activity in oculomotor brain-stem neurons sending axons directly to the eye muscles, whose activity is coordinated by interactions of nuclei commanding vertical, oblique, and lateral determinants of gaze. The vestibular neurons in the brain stem relay information about head position from inner ear to oculomotor neurons. . . . Head and eye position are related, in turn, to spinal control of posture by the reticular formation. "Without the constant and precise operation of these three systems, we could neither walk and see, nor sit still and read. . . . Together with the cerebellum, the integrated activity of these brain-stem systems is responsible for giving sighted animals complex control of their acts." Quite apart from the question of awareness, it seems that the neurological analog of sense data must go beyond irritations of sensory surfaces. In the Cartesian mode it must treat the observer's body as a part of the "external" world providing the mind with inputs to be coordinated with exteroceptive inputs by innate neurological circuitry that is fine-tuned mostly in utero and in the earliest years of extrauterine life.

From Carnap to Quine, it is ordinary thing-languages to which physicalists have looked for observation sentences, whose imputed truth values (or probability values) are to be propagated through the confirmational net by conditioning (or generalized conditioning). Quine gestures toward temporally ordered sets of triggered exteroceptors as an empirical substrate for the real epistemological action, Cartesian affirmations of ordinary observation sentences. But the proffered substrate, once mentioned, plays no further rôle in Quine's epistemology. It is anyway incapable of providing an empirical footing for his holdings true until enriched by a coördinated efferent substrate. The full-blown afferent-efferent substrate would provide a footing ("neurological solipsism") upon which holdings true and holdings probable to various degrees could supervene, but it would play no rôle, either. Bag it.

So much for the empirical side of the epistemological coin. On the other side, radical probabilism abandons Carnap's search for the fountain of rationality in a perfect ignorance prior, at the same time abandoning the idea that conditioning, or generalized conditioning, is the canonical way to change your mind. Instead, radical probabilism offers a dynamic or synchronic point of view, from which the distinction between making up your mind and changing it becomes tenuous. The Carnapian motion picture is a sequence of instantaneous frames, your successive complete probability assignments to all sentences of your language, beginning with **M** and changing every time a new conjunct is added to your data base: **M** (—), $\mathbf{M}(-|D_1)$, $\mathbf{M}(-|D_1 \& D_2)$, and so on up to your present assignment, $\mathbf{M}(-|D_1 \& D_2 \& \ldots \& D_t)$. The radical probabilist picture is less detailed in each frame, and smoother or more structural across frames in the time dimension, more like a Minkowski diagram.

Thus, making up your mind probabilistically involves making up your mind about how you will change your mind. It's not that you must map that out in fine detail, any more than you must map out your instantaneous probabilities for all sentences of your language, frame by frame. But since it no longer goes without saying that you will change your mind by conditioning or generalized conditioning (probability kinematics) any more than it goes without saying that your changes of mind will be quite spontaneous or unconsidered, these are questions about which you may make up your mind about changing your mind in specific cases. You may decide to change your mind by generalized conditioning on some set of data propositions. According to the laws of probability logic ("the probability calculus") such a decision comes to the same thing as deciding to keep your conditional probabilities on the data propositions constant when your unconditional probabilities for them change. In case your probability for one of the data propositions changes to 1, this reduces to ordinary conditioning on the data proposition you've become sure of. (Essays 3, 6, 7.)

It needs to be emphasized that becoming sure of a sentence's truth doesn't guarantee that your new conditional probabilities based on it will be the same as they were before you became sure of it. That's why Carnap required that you condition only on sentences that you regard not only as true but as recording the whole of the relevant truth that you know about. For this to imply constancy of conditional probabilities there must be available to you an infinitely nuanced assortment of data propositions to condition upon. It strikes me as a fantasy, an epistemologist's pipe-dream, to imagine that such nuanced propositions are generally accessible to us. There need be no sentence you can formulate, that fits the description "the whole of the relevant truth that you know about."[7] But the diachronic perspective of radical probabilism reveals a different dimension of nuance that you can actually use in such cases to identify a set of data propositions relative to which you expect your conditional probabilities to be unchanged by an impending observation that you think will have the effect of changing your probabilities for some of the data propositions. That will be a case where updating by probability kinematics is appropriate.

Constancy of conditional probabilities opens other options for registering and communicating the effect of experience, e.g., the option of registering the ratios ("Bayes factors") $f(A,B)$ of odds between A and B afterward and beforehand:

$$(2) \qquad f(A,B) = \frac{Q(A)/Q(B)}{P(A)/P(B)} .$$

What's conveyed by the Bayes factor is just the effect of experience, final odds with prior odds factored out. Others who accept your response to your experience, whether or not they share your prior opinion, can multiply their own prior odds between A and B by your Bayes factor to get their posterior odds, taking account of your experience.[8]

Example: Expert opinion. Jane Doe is a histopathologist who hopes to settle on one of the following diagnoses on the basis of microscopic examination of a section of tissue surgically removed from a pancreatic tumor. (To simplify matters, suppose she is sure that exactly one of the three diagnoses is correct.)

7. Given an old probability distribution, P, and a new one, Q, it is an open question whether, among the sentences D in your language for which your conditional probabilities are the same relative to Q as they are relative to P, there are any for which your new probability $Q(D)$ is 1. If so, and only then, Q can be viewed as coming from P by conditioning.

8. Such uses of Bayes factors were promoted by I. J. Good in chapter 6 of his book *Probability and the Weighing of Evidence*, Charles Griffin, London, 1950. See also *Alan Turing: The Enigma*, by Andrew Hodges, Simon and Schuster, New York, 1983, pp. 196–7. Good promotes *logarithms* of odds ("plausibilities") and of Bayes factors ("weights of evidence") as intelligence amplifiers which played a rôle in cracking the German "enigma" code during the second world war.

A = Islet cell carcinoma

B = Ductal cell carcinoma

C = Benign tumor

In the event, the experience does not drive her probability for any diagnosis to 1, but does change her probabilities for the three candidates from the following values (P) prior to the experience, to new values (Q):

	A	B	C
P	1/2	1/4	1/4
Q	1/3	1/6	1/2

Henry Roe, a clinician, accepts the pathologist's findings, i.e., he adopts, as his own, her Bayes factors between each diagnosis and some fixed hypothesis, say, C[9]:

$$f(A,C) = 1/3, \quad f(B,C) = 1/3, \quad f(C,C) = 1.$$

It is to be expected that, *given a definite diagnosis,* his conditional probabilities for the prognoses "live" (for 5 years) and "die" (within 5 years) are stable, unaffected by the pathologist's report. For definiteness, suppose those stable probabilities are as follows, where lowercase "p" and "q" are used for the clinician's prior and posterior probabilities, to distinguish them from the pathologist's.

$$q(\text{live} \mid D) = p(\text{live} \mid D) = .4, .6, .9 \text{ when } D = A, B, C$$

$$q(\text{die} \mid D) = p(\text{die} \mid D) = .6, .4, .1 \text{ when } D = A, B, C$$

Given his prior probabilities $p(D)$ for the diagnoses and his adopted Bayes factors, these conditional probabilities determine his new odds on 5-year survival.[10] It works out as follows

9. Proposed by Schwartz, W. B., Wolfe, H. J., and Pauker, S. G., "Pathology and probabilities: a new approach to interpreting and reporting biopsies," *The New England Journal of Medicine* **305**(1981)917–923.

10. See Richard Jeffrey and Michael Hendrickson, "Probabilizing pathology," *Proceedings of the Aristotelian Society,* v. 89, part 3, 1988/89: p. 217, odds kinematics.

$$(3) \quad \frac{q(\text{live})}{q(\text{die})} = \frac{p(\text{live} \mid A)\text{p}(A)\text{f}(A,C) + p(\text{live} \mid B)\text{p}(B)f(B,C)}{p(\text{die} \mid A)\text{p}(A)\text{f}(A,C) + p(\text{die} \mid B)\text{p}(B)f(B,C)}$$
$$\frac{+\; p(\text{live} \mid C)\text{p}(C)}{+\; p(\text{die} \mid C)\text{p}(C)}$$

If the clinician's prior distribution of probabilities over the three diagnoses was flat, $p(D) = 1/3$ for each diagnosis, then the imagined numbers given above raise his new odds on 5-year survival from $p(\text{live}) : p(\text{die}) = 19 : 11$ to $q(\text{live}) : q(\text{die}) = 37 : 13$, so that his probability for 5-year survival rises from 63% to 74%.

Prima facie, the task of eliciting Bayes factors looks more difficult than eliciting odds, for Bayes factors are ratios of odds.[11] For the same reason it may seem that the pathologist's Bayes factor, (posterior odds) : (prior odds), cannot be elicited if (as it may well be) she has no definite prior odds. But if her Bayes factors would be stable over a large range of prior odds, so as to be acceptable by colleagues with various prior odds, her Bayes factors are as easily elicited as her posterior odds if only she is willing and able to *adopt* definite odds prior to her observation. Thus, if she adopts even prior odds, $P(A)/P(C) = P(B)/P(C) = 1$, her Bayes factors will simply be equal to her posterior odds. But if it is only uneven priors straightforwardly based on statistical abstracts of past cases that are cogent for her, the extra arithmetic presents no problem.

The example illustrates two contrasts between the radical probabilism advocated here and the phenomenalism I have been deprecating. The less important contrast concerns the distinction between probability and certainty as basic attitudes toward *Protokollsätze*. The more important contrast concerns the status of those attitudes toward *Protokollsätze* (or toward what they report) as foundations for all of our knowledge. Here, C. I. Lewis wears the Cartesian black hat better than Carnap: "Subtract, in what we say that we see, or hear, or otherwise learn from direct experience, *all that could conceivably be mistaken;* the remainder is the given content of the experience inducing this belief. If there were no such hard kernel in experience – e.g., what we see when we think we see a deer but

---

11. In general, elicitation is a process of *drawing forth*. Here, authenticity does not require the elicited Bayes factors to have been present in the pathologist's mind before the process began; the process may well be one in which she is induced to form a judgment, *making up* her mind probabilistically.

there is no deer – then the word 'experience' would have nothing to refer to."[12]

This is the sort of empiricism dismissed above, in which the term "experience" is understood not in its ordinary sense, as the sort of thing that makes you an experienced doctor, sailor, lover, traveller, carpenter, teacher, or whatever, but in a new sense, the sensuously given, in which experience is bare phenomenology or bare irritation of sensitive surfaces. It presupposes a unitary faculty of reason, the same for all subject matter, which, added to the sensuously given, equals good judgment. The formula itself goes back much further than Descartes, e.g., to Galen: "When I take as my standard the opinion held by the most skillful and wisest physicians and the best philosophers of the past, I say: The art of healing was originally invented and discovered by the logos [reason] in conjunction with experience. And to-day also it can only be practiced excellently and done well by one who employs both of these methods."[13]

But in this formula reason is theory, and experience is gained by purges, surgery, etc., the sort of thing Hippocrates had called dire in his famous "experiment perilous" aphorism. For experience in Galen's formula, C. I. Lewis substitutes the given. Galen's formula is

Experience + reason = medical expertise.

There's a similar formula for other kinds of knowledge and technique, with "reason" and "experience" referring to other things than they do in the case of medicine.[14] But Lewis's formula is general:

The given + reason = good judgment.

Here "reason" needs to be understood as something like a successful outcome of the project to which Carnap devoted his last 25 years, of designing a satisfactory general inductive logic. For that I have no hope, for reasons given above under the headings of "blindsight" and "perception and proprioception."

---

12. *An Analysis of Knowledge and Valuation,* Open Court, La Salle, Illinois, 1946, pp. 182–183. (Lewis's emphasis.)
13. The first sentences of "On medical experience," translated by Richard Walzer, in *Three Treatises on the Nature of Science,* Hackett, Indianapolis, 1985, p. 49.
14. Like "experience," "reason" has a different sense (comprehending *theory*) in Galen's formula from what it has in C. I. Lewis's; see pp. xx–xxxi of Michael Frede's introduction to the Galen book (note 10).

10

Carnap himself was undogmatic; with high hopes for his program, he offered general, inconclusive arguments as an invitation to join in testing the idea. In fairness to him it should be noted that I haven't tried to present the case for his program here; I've used it, or a simplistic cartoon of it, as a foil for a different program ("radical probabilism") that rejects the analytical basis that I've attributed to Carnap's program, the analysis of good judgment into an a priori probability function representing reason and a propositional data base representing experience.[15]

Radical probabilism doesn't insist that probabilities be based on certainties; it can be probabilities all the way down, to the roots. Modes of judgment (in particular, probabilizing) and their attendant standards of rationality are cultural artifacts, bodies of practice modified by discovery or invention of broad features seen as grounds to stand on. It is ourselves or our fellows to whom we justify particular judgments. Radical probabilism is often faulted as uncritical, e.g., as not requiring the pathologist to justify the Bayes factors she finds cogent; "Anything goes." But probabilizing – adoption of personally cogent odds, Bayes factors, and the like, concerning some range of matters, e.g., tumors – is a subject matter–dependent *techne*, an art of judgment for which honest diligence is not enough. In practice, justification – what makes the histopathologist's personally cogent Bayes factors cogent for her colleagues as well – is a mish-mash including the sort of certification attested by her framed diploma and her reputation among relevant cognoscenti. (Relevant: the cogencies of a certified, reputable faith healer are not transferrable to me.[16]) Personal cogency may express itself in commitment to specific action (say, excision) that would be thought irresponsible if the probabilistic judgment (odds on the tumor's being benign) were much different. With probability judgment as with judgment of truth and falsity, quality varies with subject matter; handicappers and

---

15. For Carnap's program, see his essays in *Studies in Inductive Logic and Probability*, University of California Press: Berkeley, Los Angeles and London, volume 1, Rudolf Carnap and Richard Jeffrey (eds.), 1971, volume 2, Richard Jeffrey (ed.), 1980.

16. In meteorology, radically probabilistic methods of assessing and improving the quality of probabilistic judgment have been in use since the 1950s, but such techniques remain largely unknown in medicine and other areas. For an account of such methods, see Morris H. DeGroot and Stephen E. Fienberg, "Assessing probability assessors: calibration and refinement." In *Statistical Decision Theory and Related Topics*, vol. 3. New York: Academic Press, 1982.

meteorologists are mostly useless as diagnosticians.[17] Judgments are *capta,* outputs of human transducers like our histopathologist, in whose central nervous system perceptive and proprioceptive neuronic inputs somehow yield probabilistic judgments about stained cells under her microscope. Although she is far from knowing how that works, she can know *that* it works, pretty well, and how she learned to work that way – whatever that way may prove to be.

As I see it, radical probabilism delivers the central philosophical goods that logical empiricists reached for over the years in various ways. Carnap got close, I think, with his idea of a "logical" *c*-function encoding meaning relations, but I'd radicalize that probabilism twice, cashing out the idea of meaning in terms of skilled use of observations to modify our probabilistic judgments, and cashing *that* out in terms of Bayes factors. Carnap's idea of an "ignorance" prior cumulatively modified by growth of one's sentential data base is replaced by a pragmatical view of priors as carriers of current judgment, and of rational updating in the light of experience as a congeries of skills like those of the histopathologist. Described conscious experiences are especially welcome data, for by Bayes' theorem, when the new odds come from the old by conditioning on a data proposition, the Bayes factor reduces to the likelihood ratio at the right[18]:

$$\frac{Q(A)/Q(B)}{P(A)/P(B)} = \frac{P(A \mid \text{data})/P(B \mid \text{data})}{P(A)/P(B)} = \frac{P(\text{data} \mid A)}{P(\text{data} \mid B)}.$$

Among the virtues of describable experience are utility for teaching or for routinizing skills of probabilizing ("If it's purple, the odds are $7:3$ on A against B"), and for thrashing out differences of opinion in the matter. But conscious experience eluding adequate description has some of those virtues. Example: histopathological instruction of medical students using a microscope with two eyepieces and an

17. The practical framework of Bayesian decision analysis is the native ground of such probabilizing. See e.g., *Clinical Decision Analysis* by Milton C. Weinstein, Harvey V. Fineberg, et al.: W. B. Saunders Co., Philadelphia, London, Toronto, Mexico City, Rio de Janeiro, Sydney, and Tokyo, 1980.

18. The *relevance quotient* $Q(A)/P(A)$ plays the same rôle in updating probabilities that the Bayes factor plays in updating odds. Where Q comes from P by conditioning on a data proposition, $Q(A)/P(A) = P(A \mid \text{data})/P(A) = P(\text{data} \mid A)/P(\text{data})$.

arrow of light with which the instructor indicates complex features of particular cells. The discussion of blindsight and proprioception was not meant to deny that but to call attention to the considerable rôle of unconscious inputs and inputs resisting description – a rôle we can expect to be far greater than has been noted, precisely because of that unconsciousness and resistance.

In logical positivism (= logical empiricism) the move from verification to probability as a way of cashing out "meaning" goes back to the 1930s, to Reichenbach's *Wahrscheinlichkeitslehre* (Leyden, 1935) and *Experience and Prediction* (Chicago, 1938), and to Carnap's "Testability and meaning" (*Philosophy of Science* 1936, 1937).[19] My own probabilism stems from a fascinated struggle with those sources, begun in Chicago with Carnap in the late 1940s and refocused in Princeton with Hempel in the mid-1950s. I see its departures not so much as a rejection but as a further step in the development of logical empiricism, i.e., the movement, not its particular verificationist stage ca. 1929.

19. But here Carnap's notion of confirmation is not yet definitely probabilistic.

2

Valuation and acceptance of
scientific hypotheses

1. INTRODUCTION

Churchman (4), Braithwaite (1), and Rudner (7) argue from premises acceptable to many empiricists to the conclusion that ethical judgments are essentially involved in decisions as to which hypotheses should be included in the body of scientifically accepted propositions.[1] Rudner summarizes:

Now I take it that no analysis of what constitutes the method of science would be satisfactory unless it comprised some assertion to the effect that the scientist as scientist accepts or rejects hypotheses.

But if this is so then clearly the scientist as scientist does make value judgments. For, since no scientific hypothesis is ever completely verified, in accepting a hypothesis the scientist must make the decision that the evidence is *sufficiently* strong or that the probability is *sufficiently* high to warrant the acceptance of the hypothesis. Obviously our decision regarding the evidence and respecting how strong is "strong enough", is going to be a function of the *importance,* in the typically ethical sense, of making a mistake in accepting or rejecting the hypothesis (7, p. 2).

The form of this reasoning is hypothetical: *If* it is the job of the scientist to accept and reject hypotheses, *then* the scientist must make value judgments. Now I shall argue (in effect) that if scientists make value judgments, then they neither accept nor reject hypotheses. These two statements together form a *reductio ad absurdum* of the widely held view which our authors presuppose, that science consists of a body of hypotheses which, pending further evidence,

First published by R. Jeffrey, in *Philosophy of Science,* Vol. 23, No. 8, pp. 237–246. Copyright by Williams & Wilkins, 1956.

The author wishes to express his thanks to Prof. C. G. Hempel, at whose suggestion this paper was written, and to Dr. Abner Shimony, for their helpful criticism.

1. Rudner (7) has the most explicit and unqualified statement of the point of view in question. These views stem largely from recent developments in statistics, especially from the work of Abraham Wald; see (9) and references there to earlier writings.

have been *accepted* as highly enough confirmed for practical purposes ("practical" in Aristotle's sense).

In place of that picture of science I shall suggest that the activity proper to scientists is the assignment of probabilities (with respect to currently available evidence) to the hypotheses which, on the usual view, they simply accept or reject. This is not presented as a fully satisfactory position, but rather as the standpoint to which we are led when we set the Churchman–Braithwaite–Rudner arguments free from the presupposition that it is the job of the scientist as such to accept and reject hypotheses.

In the following pages we shall frequently have to speak of probabilities in connection with rational choice, and this opens us to the danger of greatly complicating our task through involvement in the dispute between conflicting theories of probability.[2] To avoid this I shall make use of the notion that these theories are conflicting explications[3] of the concept, *reasonable degree of belief*. If this is so we shall usually be able to avoid the controversy by using the subjectivistic language ("rational degree of belief" or of "confidence") which is appropriate to the *explicandum*. This device is justified if readers find the relevant statements acceptable after freely translating them into the terminologies of their preferred *explicata*.

2. BETTING AND CHOOSING

It is commonly held that although we have no certain knowledge we must often act as if probable hypotheses were known to be true. For example it might be said that when we decide to inoculate a child against polio we are accepting as certain the hypothesis that the vaccine is free from virulent polio virus. Proponents of this view speak of "accepting a hypothesis" as a sort of inductive jump from high probability to certainty.

On this account, betting is an exceptional situation. Let H be the hypothesis that the ice on Lake Carnegie is thick enough to skate on. Now if one is willing to give odds of $4 : 1$ to anyone who will bet that not-H, and if these are the longest odds one is willing to give,

---

2. For accounts of this dispute, see (2), ch. 2, (3), §1, and (5).
3. The view that a theory of probability is an explication of a vague concept in common use (the *explicandum*) by a precise concept (the *explicatum*) is due to Carnap; cf. (2), ch. 1.

one has pretty well expressed by a sort of action a degree of belief in H four times as great as one's belief in not-H. Here one is risking a good – money – which admits of degrees, so that in the bet one is able to adjust the degree of risk to the degree of belief in H. But in actually attempting to skate, degree of commitment cannot be nicely tailored to suit degree of belief: One cannot arrange to fall in only part way in case the ice breaks.

Part of the discrepancy between betting and other choosing can be disposed of immediately. Thus far we have stressed the case where it is the bettor who proposes the stakes, and indeed there is nothing comparable to this in most other choosing. When you "bet" by trusting the ice with your own weight the stakes are, say, a dunking if you lose and an afternoon's skating if you win. These stakes are fixed by nature, not by the skater. But the same arrangement is common in actual betting, e.g., at a race track, where the bettor's problem is not to propose the stakes but rather to decide whether odds offered by someone else are fair or advantageous. In such cases bettors cannot pick their exact degrees of commitment any more than skaters can.

In betting as in other choosing, rational agents act so as to maximize expectation of value. In betting, the values at stake seem especially easy to measure: The value or utility of winning is measured by the amount of money won, and the value of losing by the amount lost. But it is well known that this identification is vague and approximate. In the bet about the ice, for example, the identification of utility with money would lead one to accept as advantageous odds of $1:1$ when the ratio of degrees of belief is $4:1$. But clearly it makes a difference whether the $1:1$ in question is $\$1:\$1$ or $\$1000:\$1000$. The former would be a good bet for someone of moderate means, but the latter would not.

The usual way out of this difficulty is to specify that the stakes be small compared with the bettor's fortune, but not small enough to be boring. The importance of finding a way out is that the ratio of stakes found acceptable is a convenient measure of degree of belief. But we have seen that it is not always a reliable measure. Therefore it seems appropriate to interpret the relationship between odds and utilities in the same way we interpret the relationship between the height of a column of mercury and temperature; the one is a reliable sign of the other within a certain range, but is unreliable outside that range, where we accordingly seek other signs (e.g., alcohol thermometers below and gas thermometers above the range of reliability of mercury).

3. RATIONAL DECISION: THE BAYES CRITERION

The decision whether or not to accept a bet separates naturally into four stages; we illustrate in the case of the bet about the hypothesis (*H*) that the ice on Lake Carnegie is thick enough for skating. First we draw up a *table of stakes* indicating what will be won or lost in each of the four situations which can arise depending on whether the hypothesis is true or not, and whether the bet is accepted or not.

		Actual state of the ice	
		H	not-*H*
C	A: accept		
h	the bet	Win $1.	Lose $1.
o			
i	not-A: don't		
c	accept	Neither win nor lose.	
e	the bet		

Table of Stakes

In this case we may suppose the utilities of the stakes to be proportional to the stakes themselves, so that the *table of utilities* looks like this:

	H	not-*H*
A	1	−1
not-*A*	0	0

Here we may ignore the numbers themselves, and focus on their ratios; the same information about the utilities is contained in any table which is the same as the one above except that all entries are multiplied by some positive number, e.g., $\begin{bmatrix} 5 & -5 \\ 0 & 0 \end{bmatrix}$

If somehow we know that the bettor's degrees of belief in *H* and not-*H* stand in the ratio 4:1 we can construct a *table of expectations:*

17

	H	not-H
A	4	-1
not-A	0	0

where the expectation in each of the four situations is the product of belief in and utility of that situation.

By adding the two numbers in the top row of this table we get the *total expectation* from choice A (accepting the bet): 3. The sum of the numbers in the bottom row is the total expectation from not-A. The Bayes criterion defines the rational choice to be the one with the greater expectation. Here, then, the rational decision would be to accept the bet.

The decision about actually trying the ice is exactly parallel. Here the table of stakes is

	H	not-H
A: try to skate	Skate	Get wet
not-A: don't try	Neither skate nor get wet	

If skating and getting wet are equal and opposite goods, the utility table and the rest of the calculation is identical with that for the betting decision, and the recommendation is: Try the ice.

The Bayes criterion has generally been accepted as a satisfactory explication of "rational choice" *relative to a set of numerical utilities and degrees of belief.* If the criterion is accepted then the rationality of a decision which conforms to it can be attacked only on grounds that the degrees of belief and utilities involved are themselves unreasonable. The most influential school of thought in statistics today [1956] holds that in many cases there are no reasonable grounds for assigning probabilities to sets of hypotheses. This does not mean that in such cases the reasonable degree of belief in each hypothesis is zero, or that it is the same for all hypotheses, but simply that no numerical assignment whatever can be justified. Accordingly, statisticians have developed alternatives to Bayes' criterion, one of which (the *minimax* criterion) we shall consider in section 5.

The question of how and whether it is possible to justify the

18

assignment of numerical utilities to situations is even more difficult, and we do not propose to consider it here. But it should be noted that the use made of utilities by the Bayes criterion is not very exacting. As observed earlier, we are concerned less with the utilities themselves than with their ratios. But not even that much is required, for in applying the Bayes criterion to a choice between two actions it is sufficient to know ratios of certain *differences* between the utilities. [See *The Logic of Decision*, ch. 2.]

4. CHOICE BETWEEN HYPOTHESES: BAYES' METHOD

Meno, in the dialogue bearing his name, makes a strong objection to the Socratic concept of inquiry:

> And how will you enquire, Socrates, into that which you do not know? What will you put forth as the subject of enquiry? And if you find what you want, how will you ever know that this is the thing which you did not know? (6, Steph. 80).

In reply, Socrates undertakes the famous demonstration of how geometrical ideas can be "recollected" by an ignorant boy. But then he goes on, and apparently weakens the force of the demonstration by admitting

> . . . Some things I have said of which I am not altogether confident. But that we shall be better and braver and less helpless if we think that we ought to enquire, than we should have been if we indulged in the idle fancy that there was no knowledge and no use in seeking to know what we do not know; – that is a theme upon which I am ready to fight, in word and deed, to the utmost of my power (6, Steph. 86).

This is not mere wishful thinking, but rather part of a rational argument, in the Bayes sense of "rational." Using our previous notation, the hypothesis under consideration is (H): "Knowledge is obtainable through inquiry," and the choice is between A, the decision to inquire, and not-A. Meno had made it plausible that H is very improbable; Socrates's first reply (the demonstration of recollection) was an attempt to undermine Meno's argument directly, by showing that in fact H is more probable than Meno would have us believe. Socrates's second reply, quoted above, concedes that Meno may be right, but goes on to say that even if H *is* improbable, the utility of knowledge is so great that even when it is multiplied by a small probability, *the total expectation from inquiry (A) exceeds that from not-A.*

	H	not-*H*
A	Possibility of obtaining knowledge.	Waste of effort.
not-*A*	No knowledge obtained and no effort wasted in seeking it.	

Table of Stakes

For amusement, we might assign numerical utilities to the table of stakes: $\begin{bmatrix} 1000 & -1 \\ 0 & 0 \end{bmatrix}$. A little calculation shows that with these utilities it is rational to inquire even if the probability of *H* is as low as .001.

This pattern of argument is fairly common as a justification for faith – in God (Pascal's wager), in inquiry (Socrates), in the unity of the laws of nature (Einstein). But it should be noted that in all three cases what is meant by "faith" is not verbalized intellectual acceptance of the truth of a thesis, but rather commitment to a line of action which would be useless or even damaging if the thesis in question were false. Typically, in these cases, the thesis itself is extremely vague; but it is meaningful to those who accept it in the sense that it partly determines their activity.

To take a more precise thesis as an example, consider the problem of quality control in the manufacture of polio vaccine. A sample of the vaccine in a certain lot is tested and found to be free from active polio virus. Let us suppose that this imparts a definite probability to the hypothesis that the entire lot is good. Is this probability high enough for us rationally to accept the hypothesis?

Contrast this with a similar problem about roller skate ball bearings. Imagine that here, too, a sample has been taken and that all the bearings tested have proved satisfactory; and suppose that this evidence imparts to the hypothesis that all the bearings in the lot are good the same probability that we encountered before in the case of the vaccine. As Rudner points out, we might accept the ball bearings and yet reject the vaccine because although the probabilities are the same in the two cases, the utilities are different. If the probability were just enough to lead us to accept the bearings, we should reject the vaccine because of the graver consequences of being wrong about it.

But what determines these consequences? There is nothing in the hypothesis, "This vaccine is free from active polio virus," to tell us what the vaccine is *for,* or what would happen if the statement were accepted when false. One naturally assumes that the vaccine is intended for inoculating children, but for all we know from the hypothesis it might be intended for inoculating pet monkeys. One's confidence in the hypothesis might well be high enough to warrant inoculation of monkeys, but not of children.

The trouble is that implicitly we have been discussing a utility table with these headings

	H	not-H
Accept H		
Reject H		

but there is no way to decide what numbers should be written in the blank spaces unless we know what actions depend on the acceptance or rejection of H. Bruno de Finetti sums up the case:

> I do not deem the usual expression "to accept hypothesis H_r," to be proper. The "decision" does not really consist of this "acceptance" but in *the choice of a definite action A_r.* The connection between the action A_r and the hypothesis H_r may be very strong, say "the action A_r is that which we would choose if we knew that H_r was the true hypothesis." Nevertheless, this connection cannot turn into an identification (5, p. 219).

This fact is obscured when we consider very specialized hypotheses of the sort encountered in industrial quality control, where it is clear from the contexts, although not expressly stated in the hypotheses, what actions are in view. But the vaccine example shows that even in these cases it may be necessary to make the distinction that de Finetti urges. In the case of lawlike scientific hypotheses the distinction seems to be invariably necessary; there it is certainly meaningless to speak of *the* cost of mistaken acceptance or rejection, for by its nature a putative scientific law will be relevant in a great diversity of choice situations among which the cost of a mistake will vary greatly.

In arguing for his position Rudner concedes, "The examples I have chosen are from scientific inferences in industrial quality control. But the point is clearly general in application" (7, p. 2). Rudner seems to give his reason for this last statement later on:

I believe, of course, that an adequate rational reconstruction of the procedures of science would show that every scientific inference is properly construable as a statistical inference (i.e., as an inference from a set of characteristics of a sample of a population to a set of characteristics of the whole population (7, p. 3).

But even if analysis should show that lawlike hypotheses are like the examples from industrial quality control in being inferences from characteristics of a sample to characteristics of an entire population, they are different in the respect which is of importance here, namely their generality of application. Braithwaite and Churchman are more cautious here; they confine their remarks to statistical inferences of the ordinary sort. But we have seen that even in statistics the feasibility of blurring the distinction between accepting a hypothesis and acting upon it depends on features of the statement of the problem which are not present in every inference.

5. CHOICE BETWEEN HYPOTHESES:
MINIMAX METHOD[4]

In applying the Bayes criterion to quality control we assumed that on the basis of the relative frequency of some property in a sample of a population, definite probabilities can be assigned to the various conflicting hypotheses about the relative frequency of that property in the whole population. In general such "inverse inference" presupposes a knowledge of the *prior probabilities* (or of an *a priori probability distribution*) for the hypotheses in question. On the other hand, "direct inference" from relative frequencies in an entire population to relative frequencies in samples involves no such difficulty. Wald writes,

In many statistical problems the existence of an a priori distribution cannot be postulated, and, in those cases where the existence of an a priori distribution can be assumed, it is usually unknown to the experimenter and therefore the Bayes solution cannot be determined (9, p. 16).

4. (Added 1991) Wald's attempt to assimilate statistical hypothesis-testing to the theory of zero-sum two-person games is now of only historical interest.

For these cases Wald proposes a criterion which makes no use of inverse inference.

For simplicity we consider the case where somehow it is known that one or the other of two hypotheses, H_1 or H_2, must be true. The hypotheses assign different relative frequencies of a property P to a population. A sample consisting of only one member is drawn from the population and will be inspected for this property. Wald's problem is to choose, in advance of the inspection, an *inductive rule* which tells him which hypothesis to accept under every possible assumption as to the outcome of the inspection. In this case the choice is between four rules, since the relative frequency of P in the sample can be only 0 or 100%.

Rule 1. Accept H_1 in either case.

Rule 2. Accept H_1 in case the relative frequency of P in the sample is 0, H_2 if it is 100%.

Rule 3. Accept H_2 in case the relative frequency of P in the sample is 0, H_1 if it is 100%.

Rule 4. Accept H_2 in either case.

By direct inference one can find the conditional probabilities that each of the inductive rules will lead to the right or the wrong hypothesis on the assumption that H_1 is true, and separately on the assumption that H_2 is true. From these eight probabilities together with a knowledge of the losses (negative utilities) that would result from accepting one of the H's when in fact the other is true, one can calculate a table of risks, e.g.:

Inductive rule	Risk in using this rule in case H_1 is true	Risk in using this rule in case H_2 is true
1	7	0
2	1/2	5
3	4	2
4	10	18

23

Wald's *minimax criterion* is: Minimize the maximum risk. Here this directs us to choose the rule – 3 – for which the larger of the two risks is least.

The minimax criterion is the counsel of extreme conservatism or pessimism. Wald proves this in two ways. (i) He shows that "a minimax solution is, under some weak restrictions, a Bayes solution relative to a least favorable a priori distribution" (9, p. 18). (ii) He shows that the situation in which an experimenter uses the minimax criterion to make a decision is formally identical with the situation in which the experimenter is playing a competitive "game" with a personalized Nature in the sense that the experimenter's losses are Nature's gains. Nature plays her hand by selecting a set of prior probabilities for the hypotheses between which the experimenter must choose; being intelligent and malevolent, Nature chooses a set of probabilities which are as unfavorable as possible when viewed in the light of the negative utilities which the experimenter attaches to the acceptance of false hypotheses.

Since different experimenters make different value judgments, it would seem that in applying the minimax criterion all experimenters implicitly assume that this is the worst of all possible worlds *for them*. We might look at the matter in this way: The minimax criterion is at the pessimistic end of a continuum of criteria. At the other end of this continuum is the "minimin" criterion, which advises all experimenters to minimize minimum risk. Here all act as if this were the *best* of all possible worlds *for them*. The rules at both extremes of the continuum share the same defect: They presuppose a great sensitivity on the part of Nature to human likes and dislikes and are therefore at odds with a basic attitude which we all share, in our lucid moments.

Wald was aware of the sort of objection we have been making:

The analogy between the decision problem and a two-person game seems to be complete, except for one point. Whereas the experimenter wishes to minimize the risk . . . , we can hardly say that Nature wishes to maximize [the risk]. Nevertheless, since Nature's choice is unknown to the experimenter, it is perhaps not unreasonable for the experimenter to behave as if Nature wanted to maximize the risk (9, p. 27).

This suggests that we have overstated our case. As a general inductive rule, the minimax criterion represents an unprofitable extreme of caution; nevertheless we feel that there are conditions under which it is well to act with a maximum of caution, even though it would be unwise to follow that policy for all decisions. What we lack is an account of the conditions under which it is appropriate to use the minimax criterion.[5]

Apart from this, our previous objections to the notion of "accepting" a hypothesis apply to the minimax as well as to the Bayes criterion. Wald's procedure leads us to accept an inductive rule which, once the experiment has been made, determines one of the competing hypotheses as the "best." But this means, best for making the specific choice in question, e.g., whether to inject a child with polio vaccine from a certain lot. Among the *same* hypotheses, a different one might be best with respect to a different choice, e.g., inoculating a pet monkey. Hence both the Bayes and minimax criteria permit choice between hypotheses only with respect to a set of utilities which in turn are relative to the intended applications of the hypotheses.

6. CONCLUSION

On the Churchman–Braithwaite–Rudner view it is the task of scientists as such to accept and reject hypotheses in such a way as to maximize the expectation of good for, say, a community for which they act. On the other hand, our conclusion is that if scientists are to maximize good, they should refrain from accepting or rejecting hypotheses, for they cannot possibly do so in such a way as to optimize every decision which may be made on the basis of those hypotheses. We note that this difficulty cannot be avoided by making acceptance relative to the most stringent possible set of utilities (even if there were some way of determining what that is) because then the choice would be wrong for all less stringent sets. One cannot, by accepting or rejecting the hypothesis about the polio vaccine, do justice both to the problem of the physician who is trying to decide whether to inoculate a child, and the veterinarian who has a similar problem about a monkey. To accept or reject that hypothesis once for all is to introduce an unnecessary conflict between the interests of the physician and the veterinarian. The conflict can be resolved if the scientist either provides them both with a

5. Savage (8, ch. 13) discusses a number of other objections to the minimax criterion.

25

single probability for the hypothesis (whereupon they make their own decisions based on the utilities peculiar to their problems), or takes on the job of making a separate decision as to the acceptability of the hypothesis in each case. In any event, we conclude that it is not the business of the scientist as such, least of all of the scientist who works with lawlike hypotheses, to accept or reject hypotheses.

We seem to have been driven to the conclusion that the proper role of scientists is to provide the rational agents in the society which they represent with probabilities for the hypotheses which on the other account they simply accept or reject. There are great difficulties with this view. (i) It presupposes a satisfactory theory of probability in the sense of *degree of confirmation* for hypotheses on given evidence. (ii) Even if such a theory were available there would be great practical difficulties in using it in the way we have indicated. (iii) This account bears no resemblance to our ordinary conception of science. Books on electrodynamics, for example, simply list Maxwell's equations as laws; they do not add a degree of confirmation. These are only some of the difficulties with the probabilistic view of science.

To these, Rudner adds a very basic objection.

. . . the determination that the degree of confirmation is say, *p*, . . . which is on this view being held to be the indispensable task of the scientist *qua* scientist, is clearly nothing more than *the acceptance by the scientist of the hypothesis that the degree of confidence is p . . .* (7, p. 4).

But of course we must reply that it is no more the business of scientists to "accept" hypotheses about degrees of confidence than it is to accept hypotheses of any other sort, and for the same reasons.[6] Rudner's objection must be included as one of the weightiest under heading (i) above as a difficulty of the probabilistic view of science. These difficulties may be fatal for that theory; but they cannot save the view that the scientist, qua scientist, accepts hypotheses.

6. In Carnap's confirmation theory there at first seems to be no difficulty since it is a logical rather than a factual question, what the degree of confirmation of a given hypothesis is, with respect to certain evidence. But the difficulty may appear at a deeper level in choosing a particular c-function; this Carnap describes as a practical decision. See (3), §18.

REFERENCES

1. R. B. Braithwaite, *Scientific Explanation* (Cambridge University Press, Cambridge, 1953).
2. Rudolf Carnap, *Logical Foundations of Probability* (University of Chicago Press, Chicago, 1950).
3. Rudolf Carnap, *The Continuum of Inductive Methods* (University of Chicago Press, Chicago, 1952).
4. C. West Churchman, *Theory of Experimental Inference* (The Macmillan Co., New York, 1948).
5. Bruno De Finetti, pp. 217–225 of *Proceedings of the Second Berkeley Symposium on Mathematical Statistics and Probability*, ed. Jerzy Neyman (University of California Press, Berkeley, 1951).
6. Plato, "Meno," *The Dialogues of Plato*, tr. Benjamin Jowett (3rd ed., Random House, Inc., New York), I, 349–380.
7. Richard Rudner, "The Scientist *Qua* Scientist Makes Value Judgements," *Philosophy of Science*, vol. 20 (1953), pp. 1–6.
8. Leonard J. Savage, *The Foundations of Statistics* (John Wiley & Sons, Inc., New York, 1954).
9. Abraham Wald, *Statistical Decision Functions* (John Wiley & Sons, Inc., New York, 1950).

POSTSCRIPT (1991): CARNAP'S VOLUNTARISM

It now strikes me that 35 years ago, ending this essay, I did scant justice to the voluntarism that was a constant in Carnap's attitude from well before its expression in *Der logische Aufbau der Welt* (1928),[7] through its later expression in "Empiricism, Semantics and Ontology" (1950),[8] to his death in 1970.

Carnap's voluntarism was a humanistic version of Descartes's explanation of the truths of arithmetic as holding because God willed them: not just "Let there be light," but "Let 1 + 1 = 2" and all the rest. Carnap substituted humanity for God in this scheme; that's one way to put it, a way Carnap wouldn't have liked much, but close to the mark, I think, and usefully suggestive. Item: Descartes was stonewalling, using God's fiat to block further inquiry. It

---

7. Translated by Rolf George, *The Logical Structure of the World*, University of California Press, Berkeley and Los Angeles, 1967.
8. Reprinted in the second edition of Carnap's *Meaning and Necessity*, University of Chicago Press, Chicago, 1956.

27

is not for us to inquire why He chose 2 instead of 3. But for our own fiat the question is not what it was, but what it shall be: choice of means to our chosen ends. This fiat cannot be a whim, for this choice will be made through the public deliberations of a constitutional convention, surveying alternatives and comparing their merits. In *that* sense the choice is conventional.

Philosophically, Carnap was a social democrat; his ideals were those of the enlightenment. His persistent, central idea was: "It's high time we took charge of our own mental lives" – time to engineer our own conceptual scheme (language, theories) as best we can to serve our own purposes; time to take it back from tradition, time to dismiss Descartes's God as a distracting myth, time to accept the fact that there's nobody out there but us, to choose our purposes and concepts to serve those purposes, if indeed we are to choose those things and not simply suffer them. That's a bigger "if" than Carnap would readily acknowledge. A good part of his dispute with Quine centered on it.[9] Philosophically as well as politically Quine generally spoke as a conservative, Carnap as a socialist. For Carnap, deliberate choice of the syntax and semantics of our language was more than a possibility – it was a duty we owe ourselves as a corollary of freedom.

In his last 25 years, Carnap counted specification of c-functions among the semantical rules for languages. Choice of a language was a framework question, a practical choice that could be wise or foolish, and lucky or unlucky, but not true or false. (It will be true that we have chosen a particular framework, but that doesn't make it a true choice. Truth and falsity don't apply to choices.) If deliberation eventuates in adoption of a framework whose semantical rules specify the confirmation function c^*, then we have made it true by fiat, by convention, by reasoned choice, that "Pa" confirms "Pb" to degree $2/3$.[10]

Then Rudner's statement at the end of section 6 above is false on Carnap's view – my footnote 6 notwithstanding. On that view it is not the individual scientist who chooses the c-function; that is a social choice, a convention specifying the framework within which

9. This was also evident in their linguistic hobbies: Carnap was a dedicated learner of artificial, would-be international languages, Quine of natural, national languages.
10. Like the truth that "Pa" logically implies " $\neg$ $\neg Pa$," that's a truth *about* the adopted framework, to be expressed not in it but in a metalanguage.

28

the scientist works. Thus it is by social fiat that $c(h, e) = p$ for particular sentences h, e, and particular numbers p. The contribution of the individual experimental scientist might be to determine that the sentence e is true in fact, but that determination is by observation, not fiat.

Today I'd put Carnap's logicism to broader judgmental use. Judgmental probabilities are not generally in the mind or brain, awaiting elicitation when needed. (My judgmental probability for that hypothesis is high enough to interest me in the question of its compatibility with radical probabilism.) Of course the hypothesis poses no difficulty for Carnap, for whom $c(h, e)$ values come with the framework, and for whom elicitation of your rational judgmental probability for h is a matter of identifying an e that represents everything you're sure of relevant to h. And of course it does pose a difficulty for those subjectivists who think that probabilities are "in the mind" in some simple sense. But for radical probabilism the important question about us is not whether we have probabilities in mind, but whether we can fabricate useful probabilistic proxies for whatever it is we have in mind.[11] For radical probabilism the question is whether it's feasible and desirable for us to train ourselves to choose probabilities or odds or Bayes factors, etc., as occasions demand, for use in our practical and theoretical deliberations. The question is not whether we are natural Bayesians but whether we can *do* Bayesianism, and, if so, whether we should. For Carnap, this last is the practical question: whether, after due consideration, we will.

11. The fabricators are not generally the ultimate users, but innovators like Bruno de Finetti, who showed us how to use exchangeability (essay 7). Intelligent users mostly choose such probabilistic constructs ready-to-wear, from catalogs.

3

Probable knowledge

The central problem of epistemology is often taken to be that of explaining how we can know what we do, but the content of this problem changes from age to age with the scope of what we take ourselves to know; and philosophers who are impressed with this flux sometimes set themselves the problem of explaining how we can get along, knowing as little as we do. For knowledge is sure, and there seems to be little we can be sure of outside logic and mathematics and truths related immediately to experience. It is as if there were some propositions – that this paper is white, that two and two are four – on which we have a firm grip, while the rest, including most of the theses of science, are slippery or insubstantial or somehow inaccessible to us. Outside the realm of what we are sure of lies the puzzling region of probable knowledge – puzzling in part because the sense of the noun seems to be cancelled by that of the adjective.

The obvious move is to deny that the notion of knowledge has the importance generally attributed to it, and to try to make the concept of belief do the work that philosophers have generally assigned the grander concept. I shall argue that this is the right move.

1. A PRAGMATIC ANALYSIS OF BELIEF

(See also essay 13.) To begin, we must get clear about the relevant sense of "belief." Here I follow Ramsey: "the kind of measurement of belief with which probability is concerned is . . . a measurement of belief *qua* basis of action."[1]

Ramsey's starting point was the thought that the desirability of a

First published by R. Jeffrey, in *The Problem of Inductive Logic*, I. Lakatos, ed. Copyright by Elsevier Science Publishers, 1968.
1. Frank P. Ramsey, "Truth and probability," in *The Foundations of Mathematics and Other Logical Essays*, R. B. Braithwaite, ed., London and New York, 1931, p. 171.

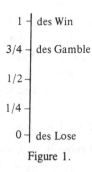

Figure 1.

gamble is a weighted average of the desirabilities of winning and losing in which the weights are the probabilities p, $1 - p$ of those outcomes. Then if these desirabilities are set at 1 and 0 as in Figure 1 the weighted average will coincide with the probability p of winning; and in any case the probability of winning will be

(1)
$$p = \frac{\text{des Gamble} - \text{des Lose}}{\text{des Win} - \text{des Lose}}.$$

Thus if the probability of winning is thought to be 3/4, the scale of desirabilities will be as in Figure 1. With desirabilities of winning and losing set at 1 and 0 the desirability of the gamble coincides with the probability 3/4 of winning; but whatever the desirabilies of winning and losing may be the gamble must lie three fourths of the way from losing to winning, so that by formula (1) the probability of winning remains p = 3/4.

On this basis Ramsey is able to give rules for deriving your subjective probability *and* desirability functions from your preference ranking of gambles, provided the preference ranking satisfies certain conditions of consistency. The probability function obtained in this way is a probability measure in the technical sense that, given any finite set of pairwise incompatible propositions which together exhaust all possibilities, their probabilities are nonnegative real numbers that add up to 1. And in an obvious sense, probability so constructed is a measure of your willingness to act on your beliefs in propositions: It is a measure of degree of belief.

I propose to use what I take to be an improvement of Ramsey's scheme, in which the work Ramsey does with the operation of forming gambles is done with the usual truth-functional operations on propositions.[2] The basic move is to restrict attention to certain "natural" gambles, in which the prize for winning is the truth of the proposition gambled upon, and the penalty for losing is the falsity of that proposition. In general, the situation in which you take yourself to be gambling on A with prize W and loss L is one in which you believe the proposition

$$G = AW \lor \bar{A}L.$$

If G is a natural gamble we have $W = A$ and $L = \bar{A}$, so that G is the necessary proposition, $G = AA \lor \bar{A}\bar{A} = T$. Now if A is a proposition which you think good (or bad) in the sense that you place it above T (or below T) in your preference ranking, we have

$$(2) \qquad prob\ A = \frac{des\ T - des\ \bar{A}}{des\ A - des\ \bar{A}},$$

corresponding to Ramsey's formula (1).

Here the basic idea is that if $A_1, A_2, \ldots, A_n$ are an exhaustive set of incompatible ways in which the proposition A can come true, the desirability of A must be a weighted average of the desirabilities of the ways in which it can come true:

$$(3) \qquad des\ A = w_1\ des\ A_1 + w_2\ des\ A_2 + \cdots + w_n\ des\ A_n.$$

where the weights are the conditional probabilities.

$$(4) \qquad w_i = prob\ A_i / prob\ A.$$

Let us call a function *des* which attributes real numbers to propositions a *Bayesian desirability function* if there is a probability measure *prob* relative to which (3) holds for all suitable A, $A_1, A_2, \ldots, A_n$. And let us call a preference ranking of propositions *coherent* if there is a Bayesian desirability function which ranks those propositions in order of magnitude exactly as they are ranked in order of preference. One can show[3] that if certain weak conditions are met by a coherent preference ranking, the underlying desirability function is determined up to a fractional linear transfor-

2. See Richard C. Jeffrey, *The Logic of Decision*, paperback, corrected 2nd ed., University of Chicago Press, 1990, the mathematical basis for which can be found in Ethan Bolker, *Functions Resembling Quotients of Measures*, Ph.D. Dissertation, Harvard University, 1965, and *Trans. Am. Math. Soc.*, **124**, 1966, pp. 293–312.
3. Jeffrey, op. cit., chs. 6, 8.

mation, i.e., if *des* and *DES* both rank propositions in order of magnitude exactly as they are ranked in order of preference, there must be real numbers a, b, c, d such that for any proposition A in the ranking we have

$$(5) \qquad DES\ A = \frac{a\ des\ A + b}{c\ des\ A + d}.$$

The probability measure *prob* is then determined by (2) up to a certain quantization. In particular, if *des* is Bayesian relative to *prob*, then *DES* will be Bayesian relative to *PROB*, where

$$(6) \qquad PROB\ A = prob\ A\ (c\ des\ A + d).$$

Under further plausible conditions, (5) and (6) are given either exactly (as in Ramsey's theory) or approximately by

$$(7) \qquad DES\ A = a\ des\ A + b,$$

$$(8) \qquad PROB\ A = prob\ A.$$

I take the principal advantage of the present theory over Ramsey's to be that here we work with your actual beliefs, whereas Ramsey needs to know what your preference ranking of relevant propositions would be if your views of what the world is were to be changed through having come to believe that various arbitrary and sometimes bizarre causal relationships had been established via gambles.[4]

To see more directly how preferences may reflect beliefs in the present system, observe that by (2) we must have *prob A* > *prob B* if the relevant portion of the preference ranking is

$$A, B$$
$$T$$
$$\bar{B}$$
$$\bar{A}$$

In particular, suppose that A and B are the propositions: Min gets job 1, Min gets job 2. Pay, working conditions, etc., are the same, so that Min ranks A and B together. Now if she thinks herself more likely to get job 1 than job 2, she will prefer a guarantee of ($\bar{B}$) not getting job 2 to a guarantee of ($\bar{A}$) not getting job 1; for she thinks that an assurance of not getting job 2 leaves her more likely to get

4. Jeffrey, op. cit., pp. 156–161.

one or the other of the equally liked jobs than would an assurance of not getting job 1.

2. PROBABILISTIC ACTS AND OBSERVATIONS

We might call a proposition *observational* for Min now if she can now make an observation of which the *direct* effect will be to change her degree of belief in the proposition to 0 or 1. Similarly, we might call a proposition *optional*[5] for her now if she can now perform an act of which the *direct* effect will be to change her degree of belief in the proposition to 0 or 1. Under ordinary circumstances, the proposition that the sun is shining is observational and the proposition that she blows her nose is optional. Performance of an act may give the agent what Anscombe calls "knowledge without observation" of the truth of an optional proposition.[6] Evidently, a proposition can be optional or observational without the agent knowing that it is; and the agent can be mistaken in thinking a proposition optional or observational.

The point and meaning of the requirement that the effect be "direct" in the definitions of "optional" and "observational" can be illustrated by the case of a sleeper, Henry, who awakens and sees that the sun is shining. Then one might take the observation to have shown him, directly, that the sun is shining, and to have shown him indirectly that it is not midnight. In general, an observation will cause numerous changes in Henry's belief function, but many of these can be construed as consequences of others. If there is a proposition E such that the *direct* effect of the observation is to change his degree of belief in E to 1, then for any proposition A in Henry's preference ranking, his degree of belief in A after the observation will be the conditional probability

$$(9) \qquad prob_E\, A = prob(A \mid E) = \frac{prob\ AE}{prob\ E}$$

where *prob* is his belief function before the observation. And conversely, if the observer's belief function after the observation is $prob_E$ and $prob_E$ is not identical with *prob*, then the *direct* effect of

---

5. This term replaces the barbarous "actual," used in the 1968 version.
6. G. E. M. Anscombe, *Intention*, §8, Oxford, 1957; 2nd ed., Ithaca, N.Y., 1963.

the observation will be to change the observer's degree of belief in E to 1.

But from a certain strict point of view, it is rarely or never that there is a proposition for which the direct effect of an observation is to change the observer's degree of belief in that proposition to 1; and from that point of view, the classes of propositions that count as observational or optional in the senses defined above are either empty or as good as empty for practical purposes. For if we care seriously to distinguish between 0.999 999 and 1.000 000 as degrees of belief, we may find that, after looking out the window, the observer's degree of belief in the proposition that the sun is shining is not quite 1, perhaps because he thinks there is one chance in a million that he is deluded or deceived in some way; and similarly for acts, where we can generally take ourselves to be at best *trying* (perhaps with very high probability of success) to make a certain proposition true.

One way in which philosophers have tried to resolve this difficulty is to postulate a phenomenalistic language in which an appropriate proposition E can always be expressed, as a report on the immediate content of experience; but for excellent reasons, this move is now in low repute.[7] The crucial point is not that 0.999 999 is so close to 1.000 000 as to make no odds, practically speaking, for situations abound in which the gap is more like one half than one millionth. Thus, in examining a piece of cloth by candlelight one might come to attribute probabilities 0.6 and 0.4 to the propositions G that the cloth is green and B that it is blue, without there being any proposition E for which the direct effect of the observation is anything near changing the observer's degree of belief in E to 1. One might think of some such proposition as that (E) *the cloth looks green or possibly blue,* but this is far too vague to yield *prob* $(G \mid E)$ = 0.6 and *prob* $(B \mid E)$ = 0.4. Certainly, there is *something* about what is seen that leads the observer to have the indicated degrees of belief in G and in B, but there is no reason to think this something expressible by a statement in the observer's language.[8] And physicalistically, there is some perfectly definite pattern of stimulation of the rods and cones of the retina which prompts the observer's belief,

---

7. See, e.g., J. L. Austin, *Sense and Sensibilia*, Oxford, 1962.
8. But see "Conditioning on Future Judgments" in essay 7.

but there is no reason to expect the observer to be able to describe that pattern or to recognize a true description of it, should it be suggested.

As Austin[9] points out, the crucial mistake is to speak seriously of the *evidence* of the senses. Even if the relevant experiences have perfectly definite characteristics by virtue of which you come to believe as you do, and by virtue of which in our example you come to have degree of belief 0.6 in *G*, it does not follow that there is a proposition *E* of which you are certain after the observation and for which you have *prob* $(G \mid E) = 0.6$, *prob* $(B \mid E) = 0.4$, etc.

In part, the quest for such phenomenological certainty seems to have been prompted by an inability to see how uncertain evidence can be used. Thus C. I. Lewis:

If anything is to be probable, then something must be certain. The data which themselves support a genuine probability, must themselves be certainties. We do have such absolute certainties, in the sense data initiating belief and in those passages of experience which later may confirm it. But neither such initial data nor such later verifying passages of experience can be phrased in the language of objective statement – because what can be so phrased is never more than probable. Our sense certainties can only be formulated by the expressive use of language, in which what is signified is a content of experience and what is asserted is the givenness of this content.[10]

But this motive for the quest is easily disposed of.[11] Thus, in the example of observation by candlelight, we may take the direct result of the observation (in a modified sense of "direct") to be that the observer's degrees of belief in *G* and *B* change to 0.6 and 0.4. Then degree of belief in any proposition *A* in the observer's preference ranking will change from *prob A* to

$$PROB\ A = 0.6\ prob\ (A \mid G) + 0.4\ prob\ (A \mid B).$$

In general, suppose that there are propositions $E_1, E_2, \ldots, E_n$, in which the observer's degrees of belief after the observation are p_1, $p_2, \ldots, p_n$; where the *E*'s are pairwise incompatible and collectively exhaustive; where for each *i*, *prob* E_i is neither 0 nor 1; and where for each proposition *A* in the preference ranking and for each

9. Austin, op. cit., ch. 10.
10. C. I. Lewis, *An Analysis of Knowledge and Valuation*, La Salle, Illinois, 1946, p. 186.
11. Jeffrey, op. cit., ch. 11.

i the conditional probability of A on E_i is unaffected by the observation:

(10) $\qquad PROB \ (A \mid E_i) = prob \ (A \mid E_i).$

Then the belief function after the observation may be taken to be $PROB$, where

(11) $\qquad PROB \ A = p_1 \, prob \ (A \mid E_1) + p_2 \, prob \ (A \mid E_2)$
$$+ \cdots + p_n \, prob \ (A \mid E_n),$$

if the observer's preference rankings before and after the observation are both coherent. Where these conditions are met, the propositions E_1, E_2, . . . , E_n, may be said to form a *basis* for the observation, the direct effect of which is to change the probabilities of propositions in the basis.

The situation is similar in the case of acts. A marksman may have a fairly definite idea of his chances of hitting a distant target, e.g., he may have degree of belief 0.3 in the proposition H that he will hit it. The basis for this belief may be his impressions of wind conditions, quality of the rifle, etc.; but there need be no reason to suppose that the marksman can express the relevant data; nor need there be any proposition E in his preference ranking in which the marksman's degree of belief changes to 1 upon deciding to fire at the target, and for which we have $prob \ (H \mid E) = 0.3$. But the pair H, $\bar{H}$ may constitute a *basis* for the act, in the sense that for any proposition A in the marksman's preference ranking, his degree of belief after his decision is

$$PROB \ A = 0.3 \, prob \ (A \mid H) + 0.7 \, prob \ (A \mid \bar{H}).$$

It is correct to describe the marksman as *trying* to hit the target; but the proposition that he is trying to hit the target cannot play the role of E above. Similarly, it was correct to describe the cloth as *looking* green or possibly blue; but the proposition that the cloth looks green or possibly blue does not satisfy the conditions for directness.

3. BELIEF: REASONS VS. CAUSES

Indeed it is desirable, where possible, to incorporate the results of observation into the structure of one's beliefs via a basis of form E,

37

$\bar{E}$ where the probability of E after the observation is nearly 1. For practical purposes, E then satisfies the conditions of directness, and the "direct" effect of the observation can be described as informing the observer of the truth of E. Where this is possible, the relevant passage of sense experience *causes* the observer to believe E; and if *prob* $(A \mid E)$ is high, belief in E may be a *reason* for believing A, and E may be spoken of as (inconclusive) *evidence* for A. But the sense experience is evidence neither for E nor for A. Nor does the situation change when we speak physicalistically in terms of patterns of irritation of our sensory surfaces, instead of in terms of sense experience: Such patterns of irritation *cause* us to believe various propositions to various degrees; and sometimes the situation can be helpfully analyzed into one in which we are caused to believe E_1, E_2, ... , E_n, to degrees $p_1, p_2, ... , p_n$, whereupon those beliefs provide *reasons* for believing other propositions to other degrees. But patterns of irritation of our sensory surfaces are not reasons or evidence for any of our beliefs, any more than irritation of the mucous membrane of the nose is a *reason* for sneezing.

When I stand blinking in bright sunlight, I can no more believe that the hour is midnight than I can fly. My degree of belief in the proposition that the sun is shining has two distinct characteristics. (a) It is 1, as close as makes no odds. (b) It is compulsory. Here I want to emphasize the second characteristic, which is most often found in conjunction with the first, but not always. Thus, if I examine a normal coin at great length, and experiment with it at length, my degree of belief in the proposition that the next toss will yield a head will have two characteristics. (a) It is $1/2$. (b) It is compulsory. In the case of the coin as in the case of the sun, I cannot decide to have a different degree of belief in the proposition, any more than I can decide to walk on air.

In my scientific and practical undertakings I must make use of such compulsory beliefs. In attempting to understand or to affect the world, I cannot escape the fact that I am part of it: I must rather make use of that fact as best I can. Now where epistemologists have spoken of observation as a source of *knowledge*, I want to speak of observation as a source of compulsory *belief* to one or another degree. I do not propose to identify a very high degree of belief with knowledge, any more than I propose to identify the property of being near 1 with the property of being compulsory.

38

Nor do I postulate any *general* positive or negative connection between the characteristic of being compulsory and the characteristic of being sound or appropriate in the light of the believer's experience. Nor, finally, do I take a compulsory belief to be necessarily a permanent one: New experience or new reflection (perhaps, prompted by the arguments of others) may loosen the bonds of compulsion, and may then establish new bonds; and the effect may be that the new state of belief is sounder than the old, or less sound.

Then why should we trust our beliefs? According to K. R. Popper,

... the decision to accept a basic statement, and to be satisfied with it, is causally connected with our experiences – especially with our *perceptual experiences*. But we do not attempt to *justify* basic statements by these experiences. Experiences can *motivate a decision,* and hence an acceptance or a rejection of a statement, but a basic statement cannot be *justified* by them – no more than by thumping the table.[12]

Now I don't understand what sort of action acceptance of a statement would be, but Popper does – and says that sometimes we decide to accept basic statements. But why does he think it impossible to justify such decisions? The answer comes right after the "table-thumping" excerpt, in a section ending with his famous swamp image of science:

It is like a building erected on piles. The piles are driven down from above into the swamp, but not down to any natural or "given" base; and when we cease our attempts to drive our piles into a deeper layer, it is not because we have reached firm ground. We simply stop when we are satisfied that they are firm enough to carry the structure, at least for the time being.[13]

The piles at their present depth are the currently accepted basic statements. Decisions to accept them, decisions not to try for more depth just now, are unjustifiable in the absence of bedrock. Therefore acceptance is gratuitous: "Basic statements are not justifiable by our immediate experiences, but are, from the logical point of view, accepted by an act, by a free decision" (p. 109). Popper sees no possibility of justification short of apodictic certainty.

To return to the question, "Why should we trust our beliefs?" one must ask what would be involved in *not* trusting one's beliefs, if belief is analyzed as in Section 1 in terms of one's preference struc-

12. K. R. Popper, *The Logic of Scientific Discovery,* London, 1959, p. 105.
13. Popper, op. cit., p. 111.

39

ture. One way of mistrusting a belief is declining to act on it, but this appears to consist merely in lowering the degree of that belief: to mistrust a partial belief is then to alter its degree to a new, more suitable value.

A more hopeful analysis of such mistrust might introduce the notion of sensitivity to further evidence or experience. Thus, Min and Hen might have the same degree of belief – 1/2 – in the proposition H_1 that the first toss of a certain coin will yield a head, but Min might have this degree of belief because she is convinced that the coin is normal, while Hen is convinced that it is either two-headed or two-tailed, he knows not which.[14] There is no question here of Hen's expressing his mistrust of the figure 1/2 by lowering or raising it, but he can express that mistrust quite handily by aspects of his belief function. Thus, if H_i is the proposition that the coin lands head up the ith time it is tossed, Hen's beliefs about the coin are accurately expressed by the function $prob_{\text{Hen}}$ where

$$prob_{\text{Hen}} \, H_i = 1/2, \, prob_{\text{Hen}} \, (H_i \mid H_j) = 1,$$

while Min's beliefs are equally accurately expressed by the function $prob_{\text{Min}}$ where

$$prob_{\text{Min}} \, (H_{i_1}, H_{i_2}, \ldots, H_{i_n}) = 2^{-n},$$

if $i_1 < i_2 < \ldots < i_n$. In an obvious sense, Min's beliefs are *firm* in the sense that she will not change them in the light of further evidence, since we have

$$prob_{\text{Min}} \, (H_{n+1} \mid H_1, H_2, \ldots, H_n) = prob_{\text{Min}} \, H_{n+1}$$
$$= 1/2,$$

while Hen's beliefs are quite tentative and in that sense, mistrusted by their holder. Still, $prob_{\text{Min}} \, H_i = prob_{\text{Hen}} \, H_i = 1/2$.

After these defensive remarks, let me say how and why I take compulsive belief to be sound, under appropriate circumstances. Bemused with syntax, the early logical positivists were chary of the notion of truth; and then, bemused with Tarski's account of truth, analytic philosophers neglected to inquire how we come to believe or disbelieve simple propositions. Quite simply put, the point is: Coming to have suitable degrees of belief in response to experience is a matter of training – a *skill* which we begin acquiring in early

14. This is a simplified version of "the paradox of ideal evidence," Popper, op. cit., pp. 407–409.

childhood, and are never quite done polishing. The skill consists not only in coming to have appropriate degrees of belief in appropriate propositions under paradigmatically good conditions of observation, but also in coming to have appropriate degrees of belief between zero and one when conditions are less than ideal.

Thus, in learning to use English color words correctly, children not only learn to acquire degree of belief 1 in the proposition that the cloth is blue, when in bright sunlight they observe a piece of cloth of hue squarely in the middle of the blue interval of the color spectrum; they also learn to acquire appropriate degrees of belief between 0 and 1 in response to observation under bad lighting conditions, and when the hue is near one or the other end of the blue region. Furthermore, understanding of the English color words will not be complete until it is understood, in effect, that blue is between green and violet in the color spectrum: Understanding of this point, or lack of it, will be evinced in the sorts of mistakes made and not made, e.g., a mistake of green for violet may evince confusion between the meanings of "blue" and of "violet," in the sense that the mistake is linguistic, not perceptual.

Clearly, the borderline between factual and linguistic error becomes cloudy, here, but in a perfectly realistic way, corresponding to the intimate connection between the ways in which we experience the world and the ways in which we speak. It is for this sort of reason that having the right language can be as important as (and can be in part identical with) having the right theory.

Then learning to use a language properly is in large part like learning such skills as riding bicycles and flying airplanes. One must train oneself to have the right sorts of responses to various sorts of experiences, where the responses are degrees of belief in propositions. This may, but need not, show itself in willingness to utter or assent to corresponding sentences. Need not, because, e.g., my cat is quite capable of showing that it thinks it is about to be fed, just as it is capable of showing what its preference ranking is, for hamburger, tuna fish, and oat meal, without saying or understanding a word. With people as with cats, evidence for belief and preference is behavioral; and speech is far from exhausting behavior.[15]

Our degrees of belief in various propositions are determined

15. Jeffrey, op. cit., pp. 68–70.

jointly by our training and our experience, in complicated ways that I cannot hope to describe. And similarly for conditional subjective probabilities, which are certain ratios of degrees of belief: To some extent, these are what they are because of our training – because we speak the languages we speak. And to this extent, conditional subjective probabilities reflect *meanings*. And in this sense, there can be a theory of degree of confirmation which is based on analysis of meanings of sentences. Confirmation theory is therefore semantical and, if you like, logical.

APPENDIX

Commenting on this paper, Patrick Suppes[16] concludes as follows.

A theory of rationality that does not take account of the specific human powers and limitations of attention may have interesting things to say, but not about human rationality.

It may be that there is no real issue between us here, but the emphasis makes me uncomfortable. In my view, the logic of partial belief is a branch of decision theory, and I take decision theory to have the same sort of relevance to human rationality that deductive logic has: The relevance is there, even though neither theory is directly about human rationality, and neither theory takes any account of the specific powers and limitations of human beings.

For definiteness, consider the following preference ranking of four sentences s, s', t, t', where s and s' are logically inconsistent, as are t and t'.

$$s$$
$$s'$$
$$t$$
$$t'$$

This ranking is *incoherent:* It violates at least one of the following two requirements. (a) Logically equivalent sentences are ranked together. (b) The disjunction of two logically incompatible sentences is ranked somewhere in the interval between them, endpoints included. Requirements (a) and (b) are part of (or anyway, implied by) a definition of "incoherent." To see that the given ranking is

16. *The Problem of Inductive Logic*, North-Holland Publishing Co., Amsterdam, 1968, pp. 186–189.

42

incoherent, notice that (a) implies that the disjunction of the sentences s, s' is ranked with the disjunction of the sentences t, t', while (b) implies that in the given ranking, the first disjunction is higher than the second. In my view, the point of classifying this ranking as incoherent is much like the point of classifying the pair s, s' as logically inconsistent: The two classifications have the same sort of relevance to human rationality. In the two cases, a rational person who made the classification would therefore decline to own the incoherent preference ranking or to believe both of the inconsistent sentences. (For simplicity I speak of belief here as an all-or-none affair.)

True enough: Since there is no effective decision procedure for logical consistency there is no routine procedure one can use for correctly classifying arbitrary rankings of sentences as incoherent or arbitrary sets of sentences as inconsistent. The relevance of incoherence and inconsistency to human rationality is rather that when we come to see that our preferences are incoherent or that our beliefs are inconsistent, we proceed to revise them. In carrying out the revision we may use decision theory or deductive logic as an aid; but neither theory fully determines how the revision shall go.

In fine, I take Bayesian decision theory to comprise a sort of *logic* of decision: The notion of coherence has much the same sort of relationship to human ideals of rationality that the notion of consistency has. But this is not to deny Suppes' point. The Bayesian theory is rather like a book of rules for chess which tells the reader what constitutes winning: There remains the question of ways and means.

4

Probability and the art
of judgment

PROBABILISM

In the middle of the seventeenth century, philosophers and mathematicians floated a new paradigm of judgment[1] that was urged upon the educated public with great success, notably in a prestigious "how to think" book, *The Port-Royal Logic* (Arnauld 1662), which stayed in print, in numerous editions, for over two centuries. We judge in order to act, and gambling at odds is the paradigm of rational action. ("Anything you do is a gamble!") That was the new view, in which judgment was thought to concern the desirabilities and probabilities of the possible outcomes of action, and canons for consistency of such judgments were seen as a logic of uncertain expectation. In time (ca. 1920), "The probability of rain is 20 percent" came to be possible as a weather forecast, the point of which is to identify 4:1 as the "betting odds" (ratio of gain to loss) at which the forecaster would think a gamble on rain barely acceptable, e.g., the gamble a farmer makes when he decides to harvest his alfalfa.[2]

Such is probabilism. *Radical* probabilism adds the "nonfounda-

First published by R. Jeffrey, in *Observation, Experiment, and Hypothesis in Modern Physical Science*, P. A. Chinstein and O. Hannaway, eds. Copyright by MIT Press, 1985.

1. This began with the famous Fermat-Pascal correspondence (1654) that Huygens reported in his widely read textbook (1657). Four high points in the subject's first century: James Bernoulli 1713, de Moivre 1718, Daniel Bernoulli 1738, Bayes 1763. Laplace supplied the "classical" philosophical (1795) and mathematical (1812) statements. The view that what emerged in 1654 was a new paradigm of judgment stems from Ramsey (1931) and de Finetti (1931).

2. Hallenbeck 1920, p. 647: "There probably are no more than 10 days during the year when fair weather can be forecast for the Pecos Valley with certainty. . . . Knowing this, the farmers of this district naturally wish to choose occasions for certain operations when the rain hazard is least. In the cutting and curing of alfalfa, most of them will accept a risk of 20 per cent – a good deal, however, depends on the state of the crop, the press of other work, etc."

tional" thought that there is no bedrock of certainty underlying our probabilistic judgments. The 20 : 80 probability balance between rain and dry need not be founded upon a certainty that the air has a certain describable feel and smell, say, and that when it feels and smells that way it rains just 20 percent of the time. Rather, probabilistic judgment may be appropriate as a direct response to experience, underived from sure judgment that the experience is of such and such a character. However, "direct" does not mean spontaneous or unschooled. Probabilizing is a skill that needs learning and polishing.

There will be more about this soon. For the moment, I only wish to identify the position briefly and locate it in relation to the antithesis between dogmatism and skepticism – an antithesis that probabilism is meant to resolve.

DOGMATISM

Descartes's dogmatism, as in this defense of the second of his *Rules for the Direction of the Mind,* is familiar:

He is no more learned who has doubts on many matters than the man who has never thought of them; nay he appears to be less learned if he has formed wrong opinions on any particulars. Hence it were better not to study at all than to occupy one's self with objects of such difficulty, that, owing to our inability to distinguish true from false, we are forced to regard the doubtful as certain; for in those matters any hope of augmenting our knowledge is exceeded by the risk of diminishing it. Thus . . . we reject all such merely probable knowledge and make it a rule to trust only what is completely known and incapable of being doubted.

I call this dogmatism because it admits of no degree of belief short of utter certainty. The contemporary weather forecaster who simply gives the probability of fair weather tomorrow as 80 percent would be seen by Descartes as not having a belief about the matter until he is prepared to state baldly that tomorrow will be fair or that it will not. In order to have what Descartes would see as opinions in such cases, we are "forced to regard the doubtful as certain." This usage is implicit in ordinary talk, where, dogmatically, "belief" refers equally to the mental attitude and its propositional object. In that usage, my belief about tomorrow's weather (if I have one) may be the proposition that it will be fair or the proposition that it will not, but it cannot be the nonpropositional judgmental balance of 4 : 1 between those possibilities.

45

By projecting probability from the mind into the world, one might try to place probabilism within the dogmatic framework. Is your judgment about tomorrow's weather describable by probability odds of 1 : 4 between rain and fair weather? If so, you fully believe the proposition that the probability of rain is 20 percent. So the move goes. It continues: If someone else thinks the odds even, then you two disagree about an objective question (i.e., "What is the real probability of rain tomorrow: 20 percent or 50 percent?"). On this view, today's probability of rain tomorrow is a physical magnitude, like today's noontime barometric pressure.

This move fails, not because the notion of real probability is bogus but because your current 1 : 4 judgment about rain versus fair weather need not represent your view about what the real probability of rain tomorrow is, now. Perhaps you lack information – say, about the current barometric pressure – in the presence of which your judgmental probability for rain would go from 20 percent to either 10 percent or 70 percent, depending on what the barometer reveals. If that is how you think matters stand, you do not think the real probability of rain is 20 percent; what you might think is "It's surely 10 percent or 70 percent, with odds of 5 : 1 on 10 percent."[3]

Meteorologists have long used percentages of probability as a way of "expressing the degree of assumed reliability of a forecast numerically" (Hallenbeck 1920, p. 645). This is clear in the earliest report on proto-probabilistic weather forecasting (Cooke 1906):

> All those whose duty it is to issue regular daily forecasts know that there are times when they feel very confident and other times when they are doubtful as to the coming weather. It seems to me that the condition of confidence or otherwise forms a very important part of the prediction, and ought to find expression. It is not fair to the forecaster that equal weight should be assigned to all his predictions and the usual method tends to retard that public confidence which all practical meteorologists desire to foster. It is more scientific and honest to be allowed occasionally to say "I feel very doubtful about the weather for to-morrow, but to the best of my belief it will be so-and-so"; and it must be satisfactory to the official and

---

3. This is because 20 splits the interval from 10 to 70 in the ratio 1 : 5. Algebraically: As your judgmental probability for rain is 20%, your probability odds ($p : 1 - p$) between the two hypotheses (10%, 70%) about the real probability must be such that $10p + 70(1 - p) = 20$. Thus, p is 5/6, and $p : 1 - p$ is 5 : 1.

useful to the public if one is allowed occasionally to say "It is practically certain that the weather will be so-and-so to-morrow."

With a view to expressing various states of doubt or certainty, as simply as possible, I now assign weights to each item of the forecast.

Cooke's scheme falls short of probabilism, because he still sees himself as making a propositional forecast, for example "rain" or "fair," with the weight indicating his confidence in the side he has chosen. Thus, he describes the third of his five weights as follows: "Very doubtful. More likely right than wrong, but probably wrong about four times out of ten."

In speaking of probability as "the degree of assumed reliability of a forecast," Hallenbeck, too, sounds as if he has stopped short of probabilism; however, it is clear from the overall context that his proposal is fully probabilistic. Cooke's "Precipitation (3)" differs from Hallenbeck's "60 percent probability of precipitation" in that Hallenbeck, but not Cooke, would regard "40 percent probability of fair weather" as the same forecast in different words. Cooke's scheme is a halfway house between dogmatism and probabilism: He first gives a quasidogmatic forecast and then qualifies it probabilistically. (On the other hand, I do not see Cooke's reference to the probable frequency of error as realistic deviationism. According to the calibration theorem given below, a probabilist should estimate that 40 percent of the propositions to which he attributes probability 60 percent will be false.)

SKEPTICISM

Descartes died four years before the start of the correspondence between Fermat and Pascal that initiated the development of probabilism in its modern, quantitative form. Like Montaigne before him, Descartes took the only alternative to dogmatism to be the utter suspension of belief that the Hellenistic epistemologists had associated with the philosopher Pyrrho. Here is Montaigne's statement of the case in his "Apology for Raymond Seybond" (1580):

I can see why the Pyrrhonian philosophers cannot express their general conception in any manner of speaking; for they would need a new language. Ours is wholly formed of affirmative propositions, which to them are utterly repugnant; so that when they say "I doubt," immediately you have them by the throat to make them admit that at least they know and are sure of this fact, that they doubt. Thus they have been constrained to take

47

refuge in this comparison from medicine, without which their attitude would be inexplicable: when they declare "I do not know" or "I doubt," they say that this proposition carries itself away with the rest, no more or no less than rhubarb, which expels evil humors and carries itself away with them.

This idea is more firmly grasped in the form of interrogation; "What do I know?" – the words I bear as a motto, inscribed over a pair of scales. [from *The Complete Essays of Montaigne*, tr. Donald M. Frame (Stanford University Press, 1957), pp. 392–393]

The motto is Montaigne's famous verbalization ("Que sçay-je?") of the Pyrrhonian shrug. The classical verbalization, known to him through the writings of Cicero and Sextus Empiricus, had been the paradoxical affirmation "I affirm nothing."

Probabilism is sometimes classified as a kind of skepticism, since the probabilist seeks to get along without dogmatic belief. And here comes the grab at the throat: "At least you know and are sure of this fact, that your probability odds between rain and fair are $1:4$." On this telling, probabilists have dogmatic beliefs about their own attitudes. Maybe so; for example, maybe you are sure that you have judgmental odds between rain and fair, and that they are $1:4$; but there is nothing in probabilism to require you to be quite clear about your own judgmental states, or even to have definite judgmental states in all cases to be clear about.

Probabilism does not say that the extreme probabilities, 0 and 1, are never to be used. Rather, it seeks to provide alternatives to such extreme judgmental states for use on those many occasions where they are inappropriate. Full dogmatic belief is represented in the probabilistic framework by odds of $1:0$ (or $2:0$, or $100:0$ – it makes no difference), and more moderate precise judgments are represented by odds such as $4:1$. But probabilism does not insist that you have a precise judgment in every case. Thus, a perfectly intelligible judgmental state is one in which you take rain to be more probable than snow and less probable than fair weather but cannot put numbers to any of the three because there is no fact of the matter. (It's not that there are numbers in your mind but it's too dark in there for you to read them.)

"THE GUIDE OF LIFE"

Cicero and Sextus had been aware of an alternative to both dogmatism and Pyrrhonism: the "probabilism" that informed the Academy during the second century B.C. under the leadership of Carneades and his successor Cleitomachus. (Some scholars hold it to be only linguistic confusion that connects this sense of probabilism with what emerged in the seventeenth century.[4]) Cicero himself was such a probabilist, but Sextus counted it as a species of dogmatism: ". . . as regards the End (or aim of life) we differ from the New Academy; for whereas the men who profess to conform to its doctrine use probability as the guide of life, we live in an undogmatic way. . . " (*Outlines of Pyrrhonism* I, p. 231). This translation echoes the familiar slogan that Bishop Butler was to introduce some fifteen centuries later at the end of this passage from the introduction to his *Analogy of Religion* (1736): ". . . nothing which is the possible object of knowledge, whether past, present, or future, can be probable to an infinite intelligence; since it cannot but be discerned absolutely as it is in itself, certainly true, or certainly false. But to Us, probability is the very guide of life."

What does this slogan mean? Butler himself took it to mean that, in the absence of conclusive reasons for belief in any of the alternatives, we should believe the most probable of them (e.g., he thought, the truth of the Christian religion). This is quite antithetical to probablism, which would have us withhold full belief in such cases. Thus, after formulating his slogan, Butler goes on to say

From these things it follows, that in questions of difficulty, or such as are thought so, where more satisfactory evidence cannot be had, or is not seen; if the result of examination be, that there appears upon the whole, any the lowest presumption on one side, though in the lowest degree greater; this determines the question, even in matters of speculation; and in matters of practice, will lay us under an absolute and formal obligation, in point of prudence and interest, to act upon that presumption or low probability, though it be so low as to leave the mind in very great doubt which is the truth. For surely *a man is really bound in prudence, to do what on the*

---

4. Cicero used the Latin *probabile* as a translation of the Greek *pithanon* (persuasive). Sixteenth-century Latin translators of Sextus followed him in this, and so English translators have used *probable* where it would be bare anarchronism to read the original as referring to probability in the sense that emerged in the mid-seventeenth century. Yet the seventeenth century was so taken with these very texts of Cicero and Sextus that it is plausible to see the putative confusion as a congenital part of the new probabilism – i.e., to see the seventeenth century as having read the new probability concept into (and, in part, out of) the "probabilism" of the second century B.C. But see Burnyeat's (forthcoming) argument that "Carneades was no probabilist."

whole appears, according to the best of his judgment, to be for his happiness, as what he certainly knows to be so. [emphasis added]

This last statement, though true, is far from justifying the adoption of dogmatic belief in the most probable alternative. On the contrary, a bet can be judged favorable even where the probability of winning is less than that of losing, but such a judgment requires joint consideration of probabilities and desirabilities; it is not enough to consider the probabilities alone (as Butler would have us do) or the desirabilities alone (as do those who buy lottery tickets simply because it would be so lovely to win). That was the point of Pascal's famous "wager," his prudential argument in favor of belief in God even if the probability of the truth of that belief is very slight. Whereas Butler urged belief in the Christian religion as the most probable of the alternatives, Pascal argued that, although sober judgment may find the odds unfavorable in that sense, that same judgment must think it worthwhile to try to believe[5] in view of the infinite desirability that belief would have in the unlikely case that God does exist.

THE GEOMETRY OF CHOICE

Perhaps the clearest general statement of the above point came at the end of the *Port-Royal Logic:* ". . . to judge what one ought to do to obtain a good or avoid an evil, one must not only consider the good and the evil in itself, but also the probability that it will or will not happen; and view geometrically the proportion that all these things have together. . . ." This geometrical view can be understood in familiar terms as a matter of balancing a seesaw. If we refer to the numerical measures of the good and the evil as their *desirabilities,* and if the course of action under consideration is thought to bestow a probability on the good that is 4 times the probability it would bestow upon the evil, then we can balance these considerations as in the seesaw of Figure 1.

The point of balance identifies the desirability of the chancy prospect of getting the good or the evil, with probabilities p and $1 -$

---

5. He suggests acting as if you believed: "taking the holy water, having masses said, etc. Even this will naturally make you believe, and deaden your acuteness." [*Pensées*, tr. F. W. Trotter (Dutton, 1958), p. 68] The wager is item 233 here, item 418 in the Krailsheimer translation.

$$\underset{e}{\rule{0pt}{0pt}} \hspace{6cm} \underset{f}{\wedge} \quad \underset{g}{\rule{0pt}{0pt}}$$

Figure 1. Positions on the seesaw represent desirabilities, weights represent probabilities, and the overall desirability of the course of action is represented by the position (f) of the fulcrum about which the opposing turning effects of the weights exactly cancel each other. As the weight (probability) at g is 4 times that at e, the position f of the fulcrum must divide the interval from e to g in the ratio $4:1$ if the seesaw is to balance. Thus, if the desirabilities are $e = 0$ and $g = 100$, the point of balance is $f = 80$.

p. The rule is that the ratio of distances from f to the good and to the evil must be the inverse of the ratio of the weights of the good and the evil. If we think of f as the desirability the gambler attributes to his situation before the outcome is determined, then the distances from f to g and to e will be the gain and the loss that the gambler faces, measured in units of desirability (as he sees it): gain $= g - f$, loss $= f - e$. If we call the ratio $(g - f):(f - e)$ of gain to loss "betting odds" and call the ratio $p:(1 - p)$ of probabilities of winning and losing the "probability odds," then the rule is this:

Probability odds are inverse to betting odds.

That is, one would as willingly take one side of the bet as the other. One regards the wager as advantageous (or disadvantageous) if the probability odds are larger (or smaller) than the inverse betting odds.[6] An equivalent statement is that the desirability f of the wager is a weighted average

$$f = pg + (1 - p)e$$

of the desirabilities of winning (g) and of losing (e), where the weights are the probabilities of winning (p) and of losing ($1 - p$).

THE WAGER

The founders of modern probability theory floated this balancing of probabilities against desirabilities as the right way to think about all

---

6. For example, advantageous: $p/(1 - p) > (f - e)/(g - f)$, or, after some algebra, $f < pg + (1 - p)e$. Since the fulcrum is to the left of the center of gravity, the seesaw inclines to the right.

worldly decisions. This is clear in *The Port-Royal Logic* (Dickoff and James translation, pp. 356–357), where the new way of thinking, given the place of honor at the end, is introduced by a discussion of the wisdom of buying lottery tickets but is quickly generalized:

These reflections may appear trifling, and indeed they are if we stop here. We may, however, turn them to very important account: Their principal use is to make us more reasonable in our hopes and fears. . . . We must enlighten those persons who take extreme and vexatious precautions for the preservation of life and health by showing that these precautions are a much greater evil than is the remote danger of the mishaps feared. We must reorient many people who conduct their lives according to maxims like these:
There is some danger in that affair.
Therefore, it is bad.
There is some advantage in this affair.
Therefore, it is good.
We ought to fear or hope for an event not solely in proportion to the advantage or disadvantage held for us but also with some consideration of the likelihood of occurrence.

The new paradigm, however, was good only for worldly deliberation:

Infinite things alone – for example, eternity and salvation – cannot be equaled by any temporal advantage. *We ought never to place them in the balance with any things of the world.* Consequently, even the slightest chance of salvation is worth more than all the goods of the world heaped together; and the slightest peril of being lost is more serious than all temporal evils, considered simply as evils. [ibid., emphasis added]

Technically, this is a better version of Pascal's wager than the one in the script his friends found after his death and published in the *Pensées* (1670). Here, but not there, we are explicitly forbidden to use the new paradigm to evaluate gambles between finitely and infinitely desirable goods. That is an essential restriction, for the new paradigm would see nothing to choose between a long shot and a sure thing when salvation is at stake. The reason is that any positive fraction of infinity, no matter how slight, is the same size as any other positive fraction and as the whole of infinity. Thus, where the fulcrum in Figure 1 is a finite distance to the right of e but infinitely far to the left of g, a single grain of probability at g will tip the scales to the right no less decisively if there are a million grains at e than if there are none there at all.

Dogmatism and probabilism are best seen as alternative forms of language, supporting different sorts of practice and different sorts of interpretation and criticism of common practices. As Montaigne suggests in the "Que sçay-je?" passage, natural languages are originally dogmatic, although there have been probabilistic accretions over the millennia (especially, in the past three centuries). From a dogmatic point of view, a decision problem is most naturally seen as as problem of what to believe; for example, a decision to leave one's umbrella at home is seen as expressing a belief that it will not rain, since that is the decision one would make if one were sure it would not rain. But, of course, it would not be an utter disaster to be caught in the rain without an umbrella, and so even dogmatists admit that the practical sort of acceptance of hypotheses (the acceptance they see as underlying a decision to leave one's umbrella at home) need not indicate certainty.

From a probabilistic point of view it is evident that the dogmatist's question of how close to unity the probability must be in order to make it reasonable to "accept" a hypothesis depends on what is at stake – on how important it is to be right. (To "accept" a hypothesis in a particular practical situation is to act as one would if one were quite sure the hypothesis were true.) Although an 80 percent probability of fair weather might warrant leaving one's umbrella at home on a particular occasion, it might not be enough to warrant sparing the expense of a canopy at a particular outdoor wedding reception. Thus, no one figure can be named that will serve as a probability just high enough for acceptance irrespective of what is at stake in the particular decision problem in question, and therefore probabilists will see practical acceptance as a bogus notion, a screen to hide the inadequacy of dogmatic decision theory. (For more about this, see Essay 2.)

This is also the point that probabilists see in Kyburg's (1961, p. 197) "lottery paradox," where we see that no number $(n - 1)/n$ can be close enough to 1 to warrant acceptance irrespective of what is at stake because, if it were, then in a fair n-ticket lottery one would accept, concerning each ticket, the hypothesis that *that* ticket will not win, since that hypothesis would have probability $(n - 1)/n;$ but one would also accept the hypothesis that *some* ticket will win, since that hypothesis would have probability 1. Thus, one would accept a logically inconsistent collection of hypotheses: (0) One of the n tickets will win, but (1) it will not be ticket 1, and (2) it will not be ticket 2, and . . . (n) it will not be ticket n. (By the way, Kyburg, who finds no fault with the notion of acceptance, takes the paradox to have a very different point, viz., that it need not be unreasonable to accept a collection of hypotheses that one knows to be logically inconsistent.)

Acceptance of that sort is thought to be a matter of belief, where "belief" ambiguously signifies both the proposition believed and the believer's attitude toward it. Conflating the two, we easily think of beliefs as presences of particular propositions in particular minds. In deciding that the butler did it, Poirot accepts a certain proposition as a guest in his mind, where it will reside alongside other such guests and numerous gatecrashers that he believes willy-nilly. Of course, one might speak in terms of representatives of propositions, e.g., sentences (French? English? Mentalese?) or, better, electrochemical configurations of the little gray cells that encode both propositional content and status as belief (rather than as hope, fantasy, or hypothesis). In any case, Poirot's present state of belief can be identified by identifying a certain bag of propositions or sentences, and Poirot's inferences can be identified via changes in the contents of the bag. One could simulate that aspect of Poirot's life by keeping a list of his beliefs in computer storage and updating it from time to time as he changes his mind. However, I find that an unattractive approach to methodology. Even if the electrochemical state of Poirot's little gray cells were eventually decoded in terms of sentences or propositions, the gesture toward neurology and computer science would be empty, I think. It is not into our heads (our "hardware" or "wetware") that I would look, but into our history and our lives, to see where the ideas of acceptance and probability come from and what they are good for. Methodologies are shared cultural artifacts – "software" in the public domain.

Before the middle of the seventeenth century, the term "probable" (Latin *probable*) meant *approvable*, and was applied in that sense,

54

univocally, to opinion and to action. A probable action or opinion was one such as sensible people would undertake or hold, in the circumstances. The writings of Descartes (d. 1650) are rich in examples of this archaic use of "probable" as *approvable*. Here is one: ". . . when it is beyond our power to discern the opinions which carry the most truth, we should follow the most probable. . . ." That is from Descartes's comments on his second maxim in the *Discourse on the Method*. He continues: "and even though we notice no greater probability in the one opinion than in the other, we at least should make up our minds to follow a particular one and afterwards consider it as no longer doubtful in its relationship to practice, but as very true and very certain. . . ."

Probability in this sense is a rough and ready notion. Thus, we are not to quibble about how we know which opinion is approvable. If there is a real question as to which side of an issue sensible people in our circumstances would take, then (in Descartes's words) "we notice no greater probability in the one opinion than in the other."

Recall what Descartes was commenting on: "My second maxim was that of being as firm and resolute in my actions as I could be, and not to follow less faithfully opinions the most dubious, when my mind was once made up regarding them, than if these had been beyond doubt." It seems that an action is to count as a case of following a particular opinion if it is such as one would perform who had no doubt of that opinion (= belief = proposition). In that sense, carrying an umbrella counts as following the opinion that it will rain, and leaving it at home counts as following the opinion that it will not rain. I suppose that to consider a doubtful opinion as "very true and very certain" in relation to practice is simply to act on that opinion no less decisively than one would have done in the absence of doubt. In such ways one can make sense of what would be dark sayings if one were to forget that Descartes was a dogmatist. He failed to see why it might be wise to "follow" the less likely of two opinions, e.g., by gambling on rain when you take the probability odds to be $1:4$, provided you see the betting odds as better than $4:1$. But that is just to say that he failed to anticipate the new paradigm of judgment that emerged in the Fermat-Pascal correspondence four years after his death.

Betting odds were an old instrument on which the founders of modern probability theory took a hopeful hold, keen to use it on everything in sight – not least the new science. New methodologies were abroad for which there were large claims: Bacon's in England, Descartes's on the continent. Perhaps the instrument that had been put only to trivial uses might prove to be a new organon for rightly conducting the reason and seeking truth in the sciences. In the foreword to his textbook *On Calculating in the Games of Luck* (1657), Huygens expressed some such hope: ". . . I would like to believe that in considering these things more attentively, the reader will soon see that the matter here is not a simple game of chance, but that we are laying the foundations of a very interesting and deep speculation." Thirty-three years later he made the claim quite explicitly and confidently, in introducing his *Treatise on Light:*

There will be seen . . . demonstrations of those kinds which do not produce as great a certitude as those of geometry, and which even differ much therefrom, since, whereas the geometers prove their propositions by fixed and incontestable principles, here the principles are verified by the conclusions to be drawn from them; the nature of these things not allowing of this being done otherwise. It is always possible thereby to attain to a degree of probability which very often is scarcely less than complete proof. To wit, when things which have been demonstrated by the principles that have been assumed correspond perfectly to the phenomena which experiment has brought under observation; especially when there are a great number of them, and further, principally, when one can imagine and foresee new phenomena which ought to follow from the hypotheses which one employs, and when one finds that therein the fact corresponds to our prevision. But if all these proofs of probability are met with in that which I propose to discuss, as it seems to me they are, this ought to be a very strong confirmation of the success of my inquiry; and it must be ill if the facts are not pretty much as I represent them.

This thought, that something like moral certainty is attainable "when things which have been demonstrated by the principles that have been assumed correspond perfectly to the phenomena which experiment has brought under observation," is no novelty of probabilism. Indeed, it is urged by Descartes in the same sort of context at the conclusion of his *Principles of Philosophy* (part IV, principle CCV): ". . . they who observe how many things regarding the magnet, fire, and the fabric of the whole world, are here deduced from a

56

very small number of principles, although they considered that I had taken up these principles at random and without good grounds, they will yet acknowledge that it could hardly happen that so much would be coherent if they were false."

What probabilism does here is explain a methodological precept that dogmatists accept, unexplained, in a not-quite-valid form, i.e., what I shall call the converse entailment condition.

Converse entailment condition. If H entails E, then truth of E confirms H.

[The terminology is designed to fit with that of Hempel (1965, p. 31).] In dogmatic terms of acceptance and rejection, the project of repairing and then justifying that precept looks unpromisingly like the project of tinkering with the following invalid rule of inference so as to get a rule that is both valid and useful:

Fallacy of affirming the consequent. If H entails E, and E is true, then H is true.

However, the project is straightforward in probabilistic terms, where we understand confirmation as improvement of the odds on truth:

Definition of confirmation. E confirms H if and only if the posterior odds on H are better than the prior odds, i.e., $p(E$ and $H):p(E$ but not $H) > p(H):p(\text{not } H)$.

Now, if H entails E (see Figure 2), the proposition E *and* H is H itself, while E *but not* H is the donut, D. Then, under the hypothesis of the converse entailment condition, the posterior odds on H will be $p(H):p(D)$, which will be greater than the prior odds $P(H):p(\text{not } H)$ as long as the common numerator, $p(H)$, is not 0 and the first denominator, $p(D)$, is less than the second, $p(\text{not } H)$ – as it must be, if $p(E)$ is not 1. Thus we have proved the validity of the following (cleaned-up) probabilistic version of the converse entailment condition:

Huygens's rule. If H implies E, then E confirms H, unless the prior

57

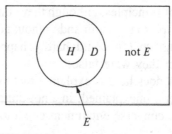

Figure 2. Proof of Huygens's rule. Areas represent probabilities. Prior odds on H are the ratio of $p(H)$ to $p(D) + p(not\ E)$, and posterior odds are the ratio of $p(H)$ to $p(D)$. If $p(H)$ and $p(not\ E)$ are both positive, the second ratio is the greater.

probability of H was 0 (H was written off as false, in advance) or that of E was 1 (E is no news).

The term "Huygens's rule" (new here) seems apt – not because that one rule exhausts Huygens's methodological remarks, quoted above, but because it represents a salient, central feature of them.

The foregoing was a case of probabilistic domestication of what was first seen as an unrationalized bit of dogmatic methodology. Before leaving the topic, let me note a similar domestication of the (rationalized) core of Karl Popper's skeptical methodology. Taking confirmation to be an idle conceit (he sees no alternative to skepticism but dogmatism), Popper holds that the aim of science can only be the refutation of false hypotheses. Therefore his central methodological rule is to test less probable hypotheses first, since lower probability means a better chance for a definitive result of the test, i.e., a refutation of the hypothesis. (Popper thinks that the converse consequence condition is a howler, viz., the fallacy of affirming the consequent.) Now it seems apt to attach Popper's name to the following probabilistic rule, which identifies the analog of Popper's basic idea (albeit within a framework that he rejects).

Popper's rule. The least probable consequences of a hypothesis are (1) the most confirmatory for it if they prove true, and (2) the most likely to refute it, by proving false. [For (1), the hypothesis must have positive prior probability.]

The comparative notion of confirmation used in (1) is to be under-

58

stood in the obvious way: that the most strongly confirmatory evidence is the evidence that gives H the greatest posterior odds.

Proof of (1). In Figure 2, if the H circle remains fixed, H's posterior probability increases as E (and, so, D) shrinks.

These two valid rules illustrate in detail what correct scientific hypothesis testing looks like from the probabilistic point of view. (For some further detail, see Jeffrey 1975 and Essay 5.) But equally important is the other side: the probabilistic critique of methodological fallacies. Here are four such fallacious rules, with counterexamples referring to a simple eight-ticket lottery.

Special consequence condition. If E confirms G, and G implies H, then E confirms H.

Counterexample. *E:* The winner is ticket 2 or 3. *G:* It is 3 or 4. *H:* It is neither 1 nor 2. Then $p\,(G \,|\, E) = 1/2 > p(G) = 1/4$, but $p(H \,|\, E) = 1/2 < p(H) = 3/4$.

Conjunction condition I. If E confirms G, and E confirms H, then E confirms G-*and*-H.

Counterexample. *E:* It is 1 or 3. *G:* 1 or 2. *H:* 3 or 4.

Conjunction condition II. If E confirms H, and F confirms H, then E-*and*-F confirms H.

Counterexample. *E:* 1 or 2. *F:* 2 or 3. *H:* 1 or 3.

Conjunction condition III. If E confirms G-*and*-H, then either E confirms G or E confirms H.

Counterexample. *G:* 1, 2, or 5. *H:* 3, 4, or 5. *E:* 5 or higher.

PROBABILITY AND FREQUENCY

Information about frequencies and averages can have a special sort of influence on probabilistic judgment: It can make judgmental

59

probabilities and expectations agree numerically with actual counts and their averages. That is what led von Mises (1928) and others to hold that probabilities are objective magnitudes, i.e., frequencies. But that position has proved surprisingly difficult to defend in detail; indeed, I think that frequentism has been shown to be an unworkable program (Essay 11). Even so, the problem remains of showing how the judgmental view of probability takes frequency data adequately into account.

It strikes me that this has been done by Bruno de Finetti (1937); I would call special attention to his understated treatment of frequencies in finite sequences in chapter 2. (This is not to deny the importance of the better-known treatment of the question in subsequent chapters, where the law of large numbers and the notion of exchangeability are used; however, chapter 2 has striking simplicity and unexpected power.) The basic fact is what I shall call the finite frequency theorem.

Finite frequency theorem. In any finite sequence of propositions, the expected number of truths is the sum of the probabilities.

This is very easily proved – so easily, indeed, that de Finetti is at pains to avoid applying to it any such grand term as "law" or "theorem."[7]

In proving the finite frequency theorem it is useful to adopt de Finetti's practice of identifying propositions with functions that assign truth values, 0 (falsity) and 1 (truth), to "possible worlds."[8] Then the function that assigns to each world the number of truths in that world among the propositions $A_1, \ldots, A_n$ will simply be the sum of separate functions[9] (propositions): $A_1 + \cdots + A_n$. By the *expectation* (or "expected value") of a function, f, of possible

7. On the other hand, the proof is no more trivial than that of Bayes's theorem, which is an immediate consequence of the definition of conditional probability.
8. See de Finetti 1972, pp. xviii–xxiv. Possible worlds (not de Finetti's term) are simply ways things might go (and might have gone), in all relevant detail. (Mathematical probabilists call them elementary events. They call propositions events.) More commonly, propositions are identified with sets of possible worlds: $A = \{w : A$ is true in $w\}$. De Finetti's way comes to the same thing if sets are identified with their indicators (i.e., functions that assign 1 to members and 0 to nonmembers).
9. This is defined as the function that assigns to each world w the sum of the values that the terms of the sum assign it.

60

worlds is meant the probability-weighted average of the different values the function can assume. If there are only finitely many possible values, say the $n + 1$ values x_k ($k = 0, \ldots, n$), this will be a sum

$$E(f) = x_0 p_0 + x_1 p_1 + \cdots + x_n p_n,$$

where p_k is the judgmental probability that f assumes the value x_k.[10] Note that if f is a proposition, then n is 1, and x_0 and x_1 are 0 and 1, respectively. It follows that $E(f)$ is p_1, i.e., that

The expectation of a proposition is its probability.

Now the finite frequency theorem takes the form

$$E(A_1 + \cdots + A_n) = p(A_1) + \cdots + p(A_n),$$

where A's are used instead of f's to emphasize that the functions are propositions, and where the p's on the right can be rewritten as E's. The proof is immediate, from the additivity of the expectation operator.[11]

If we divide both sides of this last equation by n and, on the left, apply the fact that $[E(f)]/n = E(f/n)$, we have the following theorem.

Finite relative frequency theorem. The expected relative frequency of truths among the propositions in any finite sequence is the (unweighted) average of their probabilities.

As de Finetti points out, this goes some way toward explaining why we often use observed frequencies as probabilities.

Example 1: The next trial. Knowing the number of successes on the past trials of an experiment, you need to form a judgment about the outcome of the next trial. If you judge all trials, past and next, to have the same probability of success, then the probability of success next will equal the relative frequency of past successes. (*Proof:* As

---

10. Thus, $p_k = p(\{w : f(w) = x_k\})$. If f is a proposition, then $x_k = k$ ($= 0,1$), and p_1 will be the set of worlds where f is true (i.e., the proposition itself, in the more usual representation).
11. Additivity is provable from the definition of expectation in terms of probabilities, above.

61

you know the relative frequency of successes on past trials, your estimate of that magnitude will be its actual value, and so the finite relative frequency theorem identifies that value as your judgmental probability for success next.)

In no way does this result depend on any assumption of independence of trials from one another. All that is required is that probability of success be judged constant from trial to trial. The lack of any assumption of independence is illustrated by the following example, where all propositions in the sequence are the same.

Example 2: Only constancy is assumed, not independence. Suppose the sequence is simply $\langle S,S,S \rangle$, with S = success on the first trial. The theorem still applies, for in this case the expected relative frequency of success is, by arithmetic,

$$E[(S + S + S)/3] = E(S),$$

which equals $p(S)$ since expectations of propositions are their probabilities. (That was a proof of this quirky special case of the finite relative frequency theorem.) Here the trials are as far from independent as they could be.

If the theorem is to be applied precisely as in example 1, you must not know too much about the past trials. You need to know the overall statistics – say, s successes in n trials – without knowing which particular ones were successes and which were failures (unless it happens that s is 0 or n). The reason is that if you know the truth value (0 or 1) of the kth proposition in the sequence, that number will be your judgmental probability for the proposition that the kth trial was a success, and the theorem will be applicable only if you are sure that all past trials had the same outcome, for only then can you be attributing the same probability to all trials.

Still, the theorem may be brought to bear indirectly when you know the detailed history of the trials but regard the details as unimportant.

Example 3: Irrelevant detail. You must name odds at which you will bet on the ace turning up next after a sequence of probative trials. You toss the die 100 times, writing 1 or 0 after each trial to record the outcome as ace or not. Suppose that if all you had known was the number s of past successes then your judgmental probability

62

of success would have been the same on all 101 trials, and so the theorem would have given probability $s\%$ for ace as outcome of the 101st toss. But in fact you are perfectly aware that (say) the first number you recorded was 0 and the fifth was 1, so that your judgmental probabilities are not the same for all 101 trials. The theorem is not directly applicable as in example 1. Still, it applies indirectly, for to regard the details as irrelevant is just to regard as correct the judgment about the next toss that you would have made if you had known only the overall statistics.

Note that the determination of which statistics are relevant is a probabilistic judgment (example 4), and that there will be cases where that judgment is simply negative, i.e., available statistics are seen as irrelevant (example 5).

Example 4: The odd-numbered trials have succeeded. In 100 probative trials, the successes have occurred precisely on the odd-numbered ones. Then, although the relative frequency of success has been 50%, the judgmental probability of success on the (odd) 101st trial may well be 100%, and will be if the finite frequency theorem is judged applicable in the manner of example 1 to the subsequence consisting of the odd-numbered trials.

Example 5: Too small a sample. If the number of past trials is very small (1, let us say), then there is little tendency to have the same judgmental probability for success on the next trial as on the past trial (i.e., 0 or 1). Even with a larger number of probative trials, other considerations may discredit the past statistics as a good measure of future probability, as when a questionable die is tossed six times and no aces appear. Here we are unlikely to use 0 as the probability of ace next, and there is no subsequence of past trials that will serve as in example 4.

Finally, note an application of these ideas to the evaluation of expertise.

Calibration theorem. $x\%$ is your estimate of the percentage of truths among propositions to which you assign $x\%$ probability.

The proof is immediate, from the linearity of the operator E: If you assign probability $x\%$ to each of (say) 100 propositions, then your estimate of the number of truths among them must be x, and so you must estimate that $x\%$ of them are true.

You are said to be perfectly calibrated if, for every x, precisely $x\%$ of the propositions to which you assign probability $x\%$ are true. Calibration is obviously a good thing. So is *refinement*, that is, sensitivity to differences between cases. Example: Suppose that it rains every day in September and never rains any other time. Then I would be perfectly calibrated if I attributed probability 30/365 to rain on each of the 365 days of the year. So would you be if you attributed 100% probability to rain on each day in September and 0% probability to rain on each other day. The difference between us would be one of sensitivity: maximum for you, minimum for me. [See DeGroot and Fienberg (1982) for more about this.]

THE EMERGENCE OF ODDS

Ian Hacking's *Emergence of Probability* (1975) is an attempt "to understand a quite specific event that occurred around 1660: the emergence of *our* concept of probability" (p. 9, Hacking's emphasis). Our concept: ". . . the probability emerging in the time of Pascal is essentially dual. It has to do both with stable frequencies and with degrees of belief"[12] (p. 10). Hacking holds (p. 1) that "neither of these aspects was self-consciously and deliberately apprehended by any substantial body of thinkers before the time of Pascal."[13]

However, the duality is only superficial if it is true (as I argued in the preceding section) that when probability is clearly understood as a mode of judgment the frequency aspect is thereby understood as well. If so, the specific event that occurred around 1660 was just the emergence of betting at odds as a paradigm of all action, within which probabilizing replaces believing.[14]

It is as a paradigm of judgment that probability odds were surely new in seventeenth-century Europe. I do not suggest that the very practice of betting at odds was new there in the seventeenth century; I do not imagine that at Byzantine chariot races betting was always a

12. Degrees of belief are judgmental probabilities.
13. See Garber and Zabell 1979 for a contrary argument.
14. Yet it was not until our century that judgmental probabilizing was clearly and persistently distinguished from the ambient dogmatism (Ramsey 1931; de Finetti 1931).

matter of supporters' tying their fortunes to their teams' fortunes by backing them at equal terms.[15] See what I mean: There is a great difference between the supportive sort of bet, through which you demonstrate solidarity with a cause or opinion by betting on it so as to embrace its fate, and the cagey Damon Runyon sort, where you pit your judgment against others' by fine tuning of odds. In Runyon's world, only a chump would back a team out of party loyalty. Whether he is Harry the Horse or the Chevalier de Méré, the shrewd bettor follows his judgment, not his heart. There were plenty of shrewd bettors in Paris, in the circles in which Pascal moved.

New or not, judgmental betting struck Pascal, the gentlemen of Port-Royal, Huygens, the Bernoullis, and many others as a likely paradigm for action of all kinds. But that sort of betting was probably not new to Europe, and was probably not peculiarly Indo-European either. Geertz (1973, chapter 15) gives a fascinating account of an entrenched, seemingly indigenous institution of cagey betting in Bali, where the two kinds of bets have names and have definite places around the cockpit:

. . . there are two sorts of bets, or *toh*. There is the single axial bet in the center between the principals (*toh ketengah*), and there is the cloud of peripheral ones around the ring between members of the audience (*toh kesasi*). The first is typically large; the second typically small. The first is collective, involving coalitions of bettors clustering around the owner; the second is individual, man to man. The first is a matter of deliberate, very quiet, almost furtive arrangement by the coalition members and the umpire huddled like conspirators in the center of the ring; the second is a matter of impulsive shouting, public offers, and public acceptances by the excited throng around its edges. And most curiously, . . . *where the first is always, without exception, even money, the second, equally without exception, is never such.* [Geertz's emphasis]

In Balinese terms, the native ground of the new paradigm is the periphery. Probabilism's paradigmatic act is the judgmental bet, *toh kesasi*.

RADICAL PROBABILISM

Part of our knowledge we obtain direct; and part by argument. The Theory of Probability is concerned with that part which we obtain by argument,

15. For example, with three teams in the race a supporter's betting odds will be 2 : 1 even if his probability odds are less than 1 : 2.

and it treats of the different degrees to which the results so obtained are conclusive or inconclusive. (Keynes 1921, p. 3)

I see the seventeenth-century emergence of probability as a fission of the concept of approval into judgmental probability and judgmental desirability (utility), with the second element coming into full view only lately, in work of Ramsey (1931).[16] In that same work Ramsey enunciated the further idea that I mark by calling his probabilism radical.

Ramsey's 1931 paper is in part a rejection of Keynes's (1921) view of knowledge and probability, according to which our probable knowledge is founded on certainties (i.e., truths known by direct experience). If the proposition E reports everything I know directly, and H is some doubtful hypothesis, then the probability I ought to attribute to H is determined by the logical relationship between E and H. If E logically implies that H is true, then my odds between H's truth and H's falsity ought to be $1:0$. If E logically implies that H is false, then the odds ought to be $0:1$. If E fails to determine whether H is true or false, then the odds ought to be $x:y$, where x is the a priori, "logical" probability that H and E are both true and y is the logical probability that H is false although E is true. Ramsey finds these Keynesian probabilities elusive:

. . . there really do not seem to be any such things as the probability relations he describes. He supposes that, at any rate in certain cases, they can be perceived; but speaking for myself I feel confident that this is not true. I do not perceive them . . . moreover I shrewdly suspect that others do not perceive them either, because they are able to come to so little agreement as to which of them relates any two given propositions. (Ramsey 1931, p. 161; 1990, p. 57)

But furthermore – and this is his radical probabilism – Ramsey denies that our probable knowledge need be based on certainties. This second point is independent of the first. Thus, one might modify Keynes's scheme by taking E to be merely the latest proposition that one has come to believe fully through direct experience, not the totality of all such direct knowledge over one's whole life to

16. See the essay "Truth and Probability," presented to the Moral Sciences Club of Cambridge University in 1926 and published after Ramsey died (Ramsey 1931, 1978). The fragment of Ramsey's construction that has to do with utility was reinvented by von Neumann and Morgenstern, whose 1943 book marks the beginning of any broad awareness of the very general concept of utility that we have today. It was Savage (1954) who gave Ramsey's 1931 work its due.

date. One might then take the a priori probabilities in Keynes's scheme to be simply one's judgmental probabilities as they were before one came to believe E – not mysterious "logical" probabilities. Keynes's foundationalism could still be formulated in these more modest terms, as the thesis that the only rational changes in judgmental probabilities are those prompted by fresh certainties. Here is Ramsey's rejection of that more modest foundationalism:

A third difficulty which is removed by our theory is the one which is presented to Mr Keynes' theory by the following case. I think I perceive or remember something but am not sure; this would seem to give me some ground for believing it, contrary to Mr Keynes' theory, by which the degree of belief in it which it would be rational for me to have is that given by the probability relation between the proposition in question and the things I know for certain. He cannot justify a probable belief founded not on argument but on direct inspection. In our view . . . there is no objection to such a possibility, with which Mr Keynes' method of justifying probable belief solely by relation to certain knowledge is quite unable to cope. (Ramsey 1931, p. 190; 1990, p. 86)

What Ramsey calls "a probable belief founded not on argument but on direct inspection" is what I would call direct probabilistic judgment: the probability odds as you take them to be, all things considered, but without breaking that consideration down into (a) a proposition E that you have just come to believe fully and (b) your judgmental probabilities as they were before you became certain of E. Sometimes such a breakdown is feasible, and in such cases it may be well to perform it, to obtain Keynes's "knowledge based on argument." But even then, the judgment that such a breakdown is possible will involve a direct probabilistic component.

Where considerations are broken down into (a) and (b) as above, the odds after becoming certain of E are obtained from the odds prior to that certainty by conditioning on E, i.e., by applying the following rule:

Your odds between H and *not H* after becoming certain of E are the same as your prior odds between E *and H* and E *but not H*.

Is that rule correct? Not necessarily; e.g., not if E is the information that this is an eight, H is the hypothesis that it is a heart, and the thing you saw that convinced you of E's truth was that the eight of spades was drawn. In this case your odds between E *and H* (eight of

67

hearts) and *E but not H* (eight of some other suit) were 1 : 3 before
the experience, but your odds between *H* and *not H* after the experience are 0 : 1. Evidently, the rule is correct only if *E* conveys all the
information that prompted your change of judgment. But, as
Ramsey points out, the prompt may have been an experience that
you do not have the words to convey. His example is the particular
auditory quality of whatever it was that you heard. ("Led?"
"Red?") Another example might be the visual experience that
prompts odds of 3 : 2 between Persi and Chico as originals of the
image just flashed on the screen: Facial recognition goes on in the
right hemisphere, verbalization in the left, and the lines between are
sparse.

Radical probabilism is a "nonfoundational" methodology. The
intended contrast is with dogmatic empiricism of Keynes's sort, in
which the foundations underpinning all our knowledge are truths
known by direct experience. (In this contrast the other part of
Keynes's view – the view of "logical" probabilities as the framework through which the upper stories of our knowledge are made to
rest on the foundation – is unimportant.) This foundationalism is
not peculiar to Keynes's empiricism; e.g., see Lewis 1946, p. 186:
"If anything is to be probable, then something must be certain. The
data which eventually support a genuine probability, must themselves be certainties. We do have such absolute certainties, in the
sense data initiating belief and in those passages of experience
which later may confirm it." Here Lewis accepts the foundationalist
thesis that the only rational changes in judgmental probabilities are
those prompted by fresh certainties. (His basis for that acceptance
seems to be a prior commitment to conditioning as the only rational
way of changing probability judgments in response to experience.)
Where probability is seen as a basic mode of judgment – i.e., where
probabilism is radical – that thesis loses its plausibility. It was the
felt need for a certainty to condition upon that dogmatized Lewis's
empiricism, and Keynes's.

PSYCHOLOGY AND PROBABILITY LOGIC

. . . a precise account of the nature of partial belief reveals that the laws of
probability are laws of consistency, an extension to partial beliefs of formal
logic, the logic of consistency. (Ramsey 1931; 1990, p. 78)

68

Dissatisfaction with the view of Poirot's judgmental state as representable by an assignment of truth values to the propositions or sentences he believes and disbelieves should be displaced, but not cured, by the substitution of probabilities for truth values. The new difficulty is that the ability to remember the probability of a single proposition requires a capacity to store no end of digits if the probability might be just any real number in the unit interval. What makes truth values manageable is that there are only two, and so each represents just one bit of information. If there are infinitely many values that a probability might take, each represents an infinity of bits.[17]

The most obvious way out is to make the probabilities only approximate. This amounts to increasing the possible numbers of values assignable to single propositions from Poirot's two to larger finite numbers so that the stored information stays finite. Now of course our probability judgments are often only approximate; however, if probabilizing were only or primarily a matter of assigning approximate probabilities to propositions, it would be a far cruder and less useful technique than I take it to be.

I think that we seldom have judgmental probabilities in mind for the propositions that interest us. I take it that what we do have are attitudes that can be characterized by conditions on probabilities, or (what comes to the same thing) by the sets of probability assignments that satisfy those conditions, where the members of those sets are precise and complete: Each assigns exact values to all propositions expressible in some language.[18]

Approximate probability assignments themselves are conditions on precise assignments. To say that $p(E) = 0.25$ to two decimal places of accuracy is to impose on any precise, complete assignment

17. For Carnap this is no problem: What your judgmental probability for H at any state ought to be is a conditional logical probability, a number $c(H \mid E)$ that is determined by finitely many bits of information in the shape of the conjunction E of all the *Protokollsätze* that you have been remembering, Poirot-style. But I find that doubly implausible; I can swallow neither protocol statements nor logical probabilities.

18. This idea of characterizing our probabilistic attitudes by sets is no novelty here. See, e.g., Good 1952, 1962, and Levi 1974, 1980. Note that the conditions determining such a set can involve other magnitudes beside probabilities – e.g., the condition that truth of A be preferred to truth of B, which I read as an inequality between expected utility conditionally on A and on B (Jeffrey 1965).

p the condition of assigning to E a value between $2/10$ and $3/10$. (If p violates that condition it cannot be a completion of the judgmental state in question.) But these are by no means the only cases, or the most useful, in which probabilistic judgment is a matter of adopting conditions that are satisfied by an infinity of precise, complete probability assignments. Here are some further examples, two plausible and two not.[19]

Example 1: Constant probability. As in example 1 above, you would view as unacceptable any assignment p that failed to attribute the same probability to all the propositions in the set $S = \{$success on trial 1, success on trial 2, ... $\}$.

Example 2: Exchangeability. In addition to the condition in example 1, you require that any acceptable assignment p attribute equal probabilities to conjunctions of equal numbers of members of S.

Example 3: Bernoulli trials. In addition to the condition of example 1, you require that p assign to any conjunction of members of S the product of the values it assigns to them separately. (Perhaps the trials are drawings from an urn of unknown composition, and success is a matter of drawing a black ball.)

Example 4: Independence. You take the truth or falsity of proposition E to have no evidentiary import for that of proposition H.

Example 3 is implausible because in it the set of precise, complete probability assignments is naturally parametrized by the unknown proportion r of black balls in the urn, concerning which you are likely to have views (either in the shape of a definite judgmental

---

19. The implausibility of example 3 illustrates Levi's (1980) contention that judgmental conditions must determine convex sets of probability assignments. However, as conditions on conditional probabilities and expectations can determine nonconvex sets, I doubt that convexity will do as a general restriction on sets that represent judgments. Example 4 is a case in point. There the set $\{p : p(H \mid E) = P(H)\}$ is not convex, for while it contains both q and r where

$$q(H) = q(E) = 0.5, \ q(H \text{ and } E) = 0.25$$

and

$$r(H) = r(E) = 0.3, \ r(H \text{ and } E) = 0.09,$$

it does not contain $s = (q + r)/2$, since $s(H \mid E) = 0.425 \neq s(H) = 0.4$.

70

probability distribution for r or in the shape of a set of such distributions). In the presence of any one such distribution, your judgments about trials are characterized not by the parametrized set but by a single exchangeable probability assignment: the weighted average of the parametrized assignments, with weights determined by your judgmental probability distribution for the parameter.[20] In the presence of a set of such judgmental distributions of r, your judgmental state will be represented, as in example 2, by a set of exchangeable assignments to the outcomes of trials.[21]

Probabilism does not suppose that we have particular probability assignments in mind, or that (whether we know it or not) our current states of mind are characterizable by single probability assignments, exact or approximate. Rather, probabilism characterizes our (judg)mental states by conditions on probability assignments – conditions that are typically satisfied by infinite sets of exact assignments. In these terms, revising judgment is a matter of revising each of those exact assignments, e.g., by conditioning on some certainty.

Your judgmental state will be represented by a region in the space of all exact probability assignments. Moving every point in that region has the effect of moving your judgmental state to a new region in probability space: Revision of judgment maps regions into regions. Does that sound like too much work, because it is described in terms of the individual points that make up the regions? It should not. When I wave my hand, I move all the points in it from certain positions in ordinary space to certain others, but moving all those points in this way is the same easy gesture as waving the hand.

Formal probability logic is the familiar elementary calculus of probabilities. Probability logic uses complete probability assignments just as deductive logic uses complete truth-value assignments.[22] For example, an assignment of truth (probability) values to some propositions is deductively (probabilistically) consistent if

20. For example, with the uniform distribution for r, the weighted average turns out to be the assignment Carnap called m^*.
21. Thus, in example 3, suppose you are convinced that all balls have the same color. Then the set of exchangeable assignments will be $\{qx + r(1 - x) : 0 \leq x \leq 1\}$, where $q(r)$ is the probability assignment corresponding to the hypothesis that the color is (is not) the one associated with success. [Thus, for each x in the unit interval, $p = qx + r(1 - x)$ is exchangeable, with $p(\text{success on trial } n) = x$ and $p(\text{success on } n \mid \text{success on } m) = 1$.]
22. To be more precise: The role of models in deductive logic is played by the probability models of Gaifman (1964).

and only if it is extendable to all propositions that are expressible in the language.[23]

Just as complete truth-value assignments would be suitable dogmatic judgmental states for gods, not humans, complete probability assignments could not be human probabilistic judgmental states. No more than human dogmatic judgment accepts complete truth-value assignments does human probabilistic judgment adopt complete probability assignments. Rather, our dogmatic judgment accepts propositions (i.e., sets of complete truth-value assignments) and, similarly, our probabilistic judgment adopts what one might call "probasitions" (i.e., sets of complete probability assignments). To accept a proposition (probasition) is to reject as incompatible with one's current judgment all the superhumanly complete truth-value (probability) assignments that fall outside it.

Relative to a complete probability assignment p, the degree of confirmation of H by D would be defined as[24]

$$pc(H \mid D) = p(H \mid D) - p(H).$$

This is the basic concept of probabilistic methodology, not only for superhumans whose judgmental states are complete probability assignments p but also for humans whose judgmental states are probasitions (i.e., sets of such complete assignments). Relative to any such set, D confirms H when $pc(H \mid D)$ is positive for each p in it, and D confirms H more than C confirms G when $pc(H \mid D)$ exceeds $pc(G \mid C)$ for each p in it.[25]

23. For deductive logic, this is Lindenbaum's lemma. For probability logic, see section 5.9 of de Finetti 1972 and section 3.10 of de Finetti 1974.
24. pc is what Carnap (1962, p. xvi) called D (i.e., degree of increase in firmness). Here I go along with Popper: Degree of confirmation (pc) is not a probability measure.
25. Finer-grained representations of probabilistic judgment would support subtler notions of confirmation. In note 21 above, nuance of judgment might be represented by what is formally a probability distribution for x. In general, such finer-grained representations make distributions over whole spaces of complete judgmental assignments do the work of probasitions. Degrees of confirmation of H by D would be distribution-weighted averages of the values $pc(H \mid E)$ for the various possible p's. Savage (1954, p. 58), endorsing an argument of Max Woodbury's, rejects these distributional representations on grounds that a weighted average of all complete judgmental assignments would be just one of the complete assignments in the average, but I see no reason to carry out that averaging operation here. [In Jeffrey 1983 (pp. 142–143) I mistakenly attributed the Woodbury argument to I. J. Good. As Issac Levi has pointed out to me, Good finds no fault with the sort of averaging that Savage and Woodbury reject.]

72

Probasitions corresponding to human judgmental states form sets of points that we find manageable *en bloc*. (See the remarks on hand waving four paragraphs back.) Here the parallel between the roles of propositions and probasitions in dogmatic and probabilistic methodology remains exact. Thus, if there are n "atomic" propositions whose truth values are of interest, it is unlikely that a dogmatist will be in the happy position of having a particular truth-value assignment in mind for them. (In that position, a capacity of n bits suffices to record whichever of the $N = 2^n$ assignments the dogmatist might accept.) Rather, dogmatic opinion must be expected to be represented by one of the 2^N molecular propositions that can be compounded out of the n atomic ones. However, a storage capacity of N bits would be needed to provide for all those possibilities, and, as Harman (1980) points out, a capacity of N bits is physically unattainable for fairly modest values of n (e.g., emphatically, for $n = 300$).[26] Of course, dogmatists do not provide for all those possibilities when $n = 300$; since in any one notation nearly 100% of the 2^N molecular propositions will be represented by sentences each too long to record, dogmatic acceptance can address only the remaining tiny fraction. Similarly for probasitions: We can deal with those that are easily parametrized or otherwise accessible, but those are only a tiny fraction of the mathematical possibilities. Where we can go is where there are roads, and though we can build any one of millions of new roads as needed we cannot build millions.

ACKNOWLEDGMENTS

For suggestions and criticisms that have prompted what I take to be improvements over earlier drafts, thanks are due to John Burgess, Loraine Daston, Bas van Fraassen, Alan Gibbard, Gilbert Harman, and, I fear, some I'm forgetting.

REFERENCES

Arnauld, Antoine. 1662. *Logic, or, The Art of Thinking ("The Port-Royal Logic")*. tr. J. Dickoff and P. James. Indianapolis: Bobbs-Merrill, 1964.

26. Harman 1980, p. 155. Note that the argument in Harman's section III ("Why We Don't Operate Purely Probabilistically") does not address the probasitional representation of judgment.

73

Bayes, Thomas. 1763. "An Essay Towards Solving a Problem in the Doctrine of Chances." *Philosophical Transactions of the Royal Society of London* 53.

Bernoulli, Daniel. 1738. "Specimen Theoriae Novae de Mensura Sortis" (Exposition of a New Theory of the Measurement of Risk). St. Petersburg Academy of Sciences. Translated in *Econometrica* 22 (1954): 23–36; reprinted in *Utility Theory: A Book of Readings*, ed. Alfred N. Page (New York: Wiley, 1968).

Bernoulli, James. 1713. *Ars Conjectandi*. Basel.

Burnyeat, Miles. Forthcoming. "Carneades Was No Probabilist." *Riverside Studies in Ancient Skepticism*.

Butler, Joseph. 1736. *The Analogy of Religion, Natural and Unrevealed, to the Constitution and Course of Nature*. London.

Carnap, Rudolf. 1950, 1962. *Logical Foundations of Probability*. University of Chicago Press.

Cicero. *De Natura Deorum* and *Academica*. Loeb Classical Library. There is a good Penguin edition of the first and an excellent out-of-print translation of the second [*The Academics of Cicero*, tr. James S. Reid (London: Macmillan, 1880)].

Cooke, W. Ernest. 1906. "Forecasts and Verifications in Western Australia." *Monthly Weather Review* 34: 23–24.

de Finetti, Bruno. 1931. "Sul significato soggetivo della probabilita" (On the Subjective Significance of Probability). *Fundamenta Mathematica* 17: 298–329.

de Finetti, Bruno. 1937. "La prévision: ses lois logiques, ses sources subjectives." Annales de l'Institut Henri Poincaré 7: 1–68. Translated in *Studies in Subjective Probability*, ed. H. E. Kyburg, Jr., and H. E. Smokler (Huntington, N.Y.: Krieger, 1980).

de Finetti, Bruno. 1972. *Probability, Induction, and Statistics*. New York: Wiley.

de Finetti, Bruno. 1970. *Teoria delle Probabilita*. Torino: Giuli Einaudi. Translated as *Theory of Probability* (New York: Wiley, 1974, 1975).

DeGroot, Morris H., and Stephen E. Fienberg. 1982. "Assessing Probability Assessors: Calibration and Refinement." In *Statistical Decision Theory and Related Topics*, vol. 3. New York: Academic.

De Moivre, Abraham. 1718, 1738, 1756. *The Doctrine of Chances*. Reprints. New York: Chelsea, 1967.

Descartes, René. ca. 1628. *Rules for the Direction of the Mind*.

Descartes, René. 1637. *Discourse on the Method of Rightly Conducting the Reason and Seeking Truth in the Sciences*. In *Philosophical Works of Descartes*, ed. E. S. Haldane and R. R. T. Ross (New York: Dover, 1955).

Fermat, Pierre de. Correspondence with Pascal. See F. N. David, *Games, Gods, and Gambling* (London: Griffin, 1962).

Gaifman, Haim. 1964. "Concerning Measures in First-Order Calculi." *Israel Journal of Mathematics* 2: 1–18.

74

Garber, Daniel, and Sandy Zabell. 1979. "On the Emergence of Probability." *Archive for History of Exact Sciences* 21: 33–53.

Geertz, Clifford. 1973. *The Interpretation of Cultures.* New York: Basic.

Good, I. J. 1952. "Rational Decisions." *Journal of the Royal Statistical Society* B 14: 107–114.

Good, I. J. 1962. "Subjective Probability as the Measure of a Non-Measurable Set." In *Logic, Methodology, and Philosophy of Science*, ed. E. Nagel et al. Stanford University Press.

Hacking, Ian. 1975. *The Emergence of Probability.* Cambridge University Press.

Hallenbeck, Cleve. 1920. "Forecasting Precipitation in Percentages of Probability." *Monthly Weather Review* 48: 645–647.

Harman, Gilbert. 1980. "Reasoning and Explanatory Coherence." *American Philosophical Quarterly* 17: 151–157.

Hempel, Carl G. 1965. *Aspects of Scientific Explanation.* New York: Free Press.

Huygens, Christiaan. 1957. *De Ratiociniis in Aleae Ludo* (On Calculating in Games of Luck). Reprinted in Huygens, *Oeuvres Completes* (The Hague: Martinus Nijhoff, 1920).

Huygens, Christiaan. 1690. *Treatise on Light.* Translation: New York, Dover, 1962.

Jeffrey, Richard C. 1956. "Valuation and Acceptance of Scientific Hypotheses." *Philosophy of Science* 23: 237–246.

Jeffrey, Richard C. 1965. *The Logic of Decision.* Second edition, revised: University of Chicago Press, 1983, 1990.

Jeffrey, Richard C. 1975. "Probability and Falsification: Critique of the Popper Program." *Synthese* 30: 95–117, 149–157.

Jeffrey, Richard C. 1977. "Mises Redux." In *Basic Problems in Methodology and Linguistics*, ed. R. E. Botts and J. Hintikka (Dordrecht: Reidel).

Jeffrey, Richard C. 1983. "Bayesianism with a Human Face." In *Testing Scientific Theories*, ed. John Earman. University of Minnesota Press.

Keynes, John Maynard. 1921. *A Treatise on Probability.* London: Macmillan.

Kyburg, Henry E., Jr. 1961. *Probability and the Logic of Rational Belief.* Middletown, Conn.: Wesleyan University Press.

Laplace, Pierre Simon, Marquis de. 1795. *Essaie philosophique sur les probabilités.* Translated as *A Philosophical Essay on Probabilities* (New York: Dover, 1951).

Laplace, Pierre Simon, Marquis de. 1812, 1814, 1820. *Theorie analytique des probabilités.*

Levi, Isaac. 1974. "On Indeterminate Probabilities." *Journal of Philosophy* 71: 391–418.

Levi, Isaac. 1980. *The Enterprise of Knowledge.* Cambridge, Mass.: MIT Press.

Lewis, Clarence Irving. 1946. *An Analysis of Knowledge and Valuation.* La Salle, Ill.: Open Court.

Pascal, Blaise. 1670. *Pensées.* Tr. A. J. Krailsheimer (New York: Penguin, 1966); F. W. Trotter (New York: Dutton, 1958). See also Fermat reference, above.

Pyrrho et al. For background on Hellenistic epistemology, see David Sedley, "The Protagonists," in *Doubt and Dogmatism,* ed. M. Schofield, M. Burnyeat, and J. Barnes (Oxford: Clarendon, 1980).

Ramsey, Frank Plumpton. 1931. "Truth and Probability." In *The Foundations of Mathematics* (London: Kegan Paul); also in *Philosophical Papers,* ed. D. H. Mellor (Cambridge University Press, 1990).

Savage, L. J. 1954. *The Foundations of Statistics.* New York: Wiley.

Sextus Empiricus. See Loeb Classical Library, no. 273 (*Outlines of Pyrrhonism*) and no. 291 (*Against the Logicians*).

von Mises, Richard. 1928. *Probability, Statistics and Truth.* Published in German. Second revised English edition: London: Allen & Unwin, 1957.

5

Bayesianism with a human face

WHAT'S A BAYESIAN?

Well, I'm one, for example. But not according to Clark Glymour (1980, pp. 68–69) and some other definers of Bayesianism and personalism, such as Ian Hacking (1967, p. 314) and Isaac Levi (1980, p. xiv). Thus it behooves me to give an explicit account of the species of Bayesianism I espouse (sections 1 and 2) before adding my bit (section 3, with lots of help from my friends) to Daniel Garber's (1983) treatment of the problem of new explanation of common knowledge: the so-called problem of old evidence.

With Clark Glymour, I take there to be identifiable canons of good thinking that get used on a large scale in scientific inquiry at its best; but unlike him, I take Bayesianism (what *I* call "Bayesianism") to do a splendid job of validating the valid ones and appropriately restricting the invalid ones among the commonly cited methodological rules. With Daniel Garber, I think that bootstrapping does well, too – when applied with a tact of which Bayesianism can give an account. But my aim here is to elaborate and defend Bayesianism (of a certain sort), not to attack bootstrapping. Perhaps the main novelty is the further rounding-out in section 3 (by John Etchemendy, David Lewis, Calvin Normore, and me) of Daniel Garber's treatment of what I have always seen as the really troubling one of Clark Glymour's strictures against Bayesianism. After that there is a coda (section 4) in which I try to display and explain how probability logic does so much more than truth-value logic.

First published by R. Jeffrey, in *Testing Scientific Theories*, J. Earman, ed. Minnesota Studies in the Philosophy of Science series, Vol. X. Copyright by University of Minnesota, 1983.

1. RESPONSE TO NEW EVIDENCE

In Clark Glymour's book, you aren't a Bayesian unless you update your personal probabilities by conditioning (a.k.a. "conditionalization"), i.e., like this:

As new evidence accumulates, the probability of a proposition changes according to Bayes' rule: the posterior probability of a hypothesis on the new evidence is equal to the prior conditional probability of the hypothesis on the evidence. (p. 69)

That's one way to use the term "Bayesian," but on that usage I'm no Bayesian. My sort of Bayesianism gets its name from another sense of the term "Bayes' rule," equally apt, but stemming from decision theory, not probability theory proper. Whereas Bayes' rule in Glymour's sense prescribes conditioning as the way to update personal probabilities, Bayes's rule in my sense prescribes what Wald (1950) called "Bayes solutions" to decision problems, i.e., solutions that maximize expected utility relative to some underlying probability assignment to the states of nature. (No Bayesian himself, Wald contributed to the credentials of decision-theoretic Bayesianism by proving that the Bayes solutions form a complete class.) The Reverend Thomas Bayes was both kinds of Bayesian. And of course, he was a third kind of Bayesian, too: a believer in a third sort of Bayes's rule, according to which the right probability function to start with is m^* [as Carnap (1945) was to call it].

Why am I not a Bayesian in Glymour's sense? This question is best answered by way of another: What is the "new evidence" on which we are to condition? (Remember: The senses are not telegraph lines on which the external world sends observation sentences for us to condition upon.) Not just any proposition that newly has probability one will do, for there may well be many of these, relative to which conditioning will yield various posterior probability distributions when applied to the prior.

All right, then: What about the conjunction of all propositions that newly have probability one? That will be the total new evidence, won't it? Why not take the kinematical version of Bayes' rule to prescribe conditioning on that total?

I answer this question in chapter 11 of my book (1965, 1983), and in a few other places (1968, 1970, 1975). In a nutshell, the answer is that much of the time we are unable to formulate any sentence upon

78

which we are prepared to condition, and in particular, the conjunction of all the sentences that newly have probability one will be found to leave too much out for it to serve as the Archimedean point about which we can move our probabilities in a satisfactory way. Some of the cases in which conditioning won't do are characterized by Ramsey (1931, "Truth and Probability," end of section 5) as follows:

I think I perceive or remember something but am not sure; this would seem to give me some ground for believing it, contrary to Mr. Keynes' theory, by which the degree of belief in it which it would be rational for me to have is that given by the probability relation between the proposition in question and the things I know for certain.

Another sort of example is suggested by Diaconis and Zabell (1982): A record of someone reading Shakespeare is about to be played. Since you are sure that the reader is either Olivier or Gielgud, but uncertain which, your prior probabilities for the two hypotheses are nearly equal. But now comes fresh evidence, i.e., the sound of the reader's voice when the record is played. As soon as you hear that, you are pretty sure it's Gielgud, and the prior value ca. .5 is replaced by a posterior value ca. .9, say. But, although it was definite features of what you heard that rightly made you think it very likely to have been Gielgud, you cannot describe those features in an observation sentence in which you now have full belief, nor would you be able to recognize such a sentence (immensely long) if someone else were to produce it.

Perhaps it is the fact that there surely is definite evidence that prompts and justifies the probability shift in the Olivier/Gielgud case, that makes some people think there must be an evidence *sentence* (observation sentence) that will yield the new belief function via conditionalization. Surely it is all to the good to be able to say just what it was about what you heard that made you pretty sure it was Gielgud. But few would be able to do that; nor is such inability a mark of irrationality; nor need one be able to do that in order to count as having had good reason to be pretty sure it was Gielgud. The Olivier/Gielgud case is typical of our most familiar sorts of updating, as when we recognize friends' faces or voices or handwritings pretty surely, and when we recognize familiar foods pretty surely by their look, smell, taste, feel, and heft.

Of course conditioning is sometimes appropriate. When? I mean, if your old and new belief functions are p and q, respectively, when is q of form p_E for some E to which p assigns a positive value? [Definition: $p_E(H)$ is the conditional probability of H on E, i.e., $p(H \mid E)$, i.e., $p(HE)/p(E)$.]

Here is an answer to the question:

(C) If $p(E)$ and $q(E)$ are both positive, then the conditions (a) $q_E = p_E$ and (b) $q(E) = 1$ are jointly necessary and sufficient for (c) $q = p_E$.

You can prove that assertion on the back of an envelope, via the Kolmogorov axioms and the definition of conditional probability. Here is a rough-and-ready verbal summary of (C):

Conditioning is the right way to update your probability judgments iff the proposition conditioned upon is not only (b) one you now fully believe, but is also (a) one whose relevance to each proposition is unchanged by the updating.

The point of view is one in which we take as given the old and new probability functions, p and q, and then ask whether the condition (c) $q = p_E$ is consistent with static coherence, i.e., the Kolmogorov axioms together with the definition of conditional probability applied to p and q separately. In (C), (a) is the ghost of the defunct condition of total evidence.

In the Olivier/Gielgud example, and others of that ilk, fresh evidence justifies a change from p to q even though $q \neq p_E$ for all E in the domain of p. What is the change, and when it is justified? Here is the answer, which you can verify on the back of the same envelope you used for (C):

(K) If "E" ranges over some partitioning of a proposition of p-measure 1 into propositions of positive p-measure, then the ("rigidity") condition
(r) $q_E = p_E$ for all E in the partitioning
is necessary and sufficient for q to be related to p by the following ("kinematical") formula:
(k) $q = \Sigma_E \, q(E) p_E$.

There is no more question of justifying (k) in (K) than there was of justifying (c) in (C): Neither is always right. But just as (C) gives necessary and sufficient conditions (a) and (b) for (c) to be right, so (K) gives (r), i.e., the holding of (a) for each E in the partitioning, as necessary and sufficient for (k) to be correct – where in each case, correctness is just a matter of static coherence of p and q separately. We know when (k) is right:

The kinematical scheme (k) yields the correct updating iff the relevance of each member E of the partitioning to each proposition H is the same after the updating as it was before.

It is an important discovery (see May and Harper 1976; Williams 1980; Diaconis and Zabell 1982) that in one or another sense of "close," (k) yields a measure q that is closest to p among those that satisfy the rigidity condition (r) and assign the new probabilities $q(E)$ to the Es, and that (c) yields a measure that is closest to p among those that satisfy the conditions (a) and (b) in (C). But what we thereby discover is that (so far, anyway) we have adequate concepts of closeness: We already knew that (k) was equivalent to (r), and that (c) was equivalent to (a) and (b) in (C). This is not to deny the interest of such minimum-change principles, but rather to emphasize that their importance lies not in their serving to justify (c) and (k) – for they don't – but in the further kinematical principles they suggest in cases where (k) holds for no interesting partitioning. To repeat: (c) and (k) are justified by considerations of mere coherence, where their proper conditions of applicability are met, i.e., (a) and (b) for (c), and (r) for (k). And where those conditions fail, the corresponding rules are unjustifiable.

Observe that in a purely formal sense, condition (r) is very weak; e.g., it holds whenever the Boolean algebra on which p and q are defined has atoms whose p-values sum to 1. [Proof: with "E" in (r) ranging over the atoms that have positive p-measure, $p(H \mid E)$ and q $(H \mid E)$ will both be 1 or both be 0, depending on whether E implies H or $-H$.] Then in particular, (k) is always applicable in a finite probability space, formally. But if (k) is to be useful to a human probability assessor, the E partitioning must be coarser than the atomistic one. To use the atomistic partitioning is simply to start over from scratch.

The Olivier/Gielgud example is one in which the partitioning is quite manageable: $\{O, G\}$, say, with O as the proposition that the reader is Olivier, and G for Gielgud. The hypothesis H that the reader (whoever he may be) married Vivien Leigh serves to illustrate the rigidity conditions. Applied to H, (r) yields

$$q(H \mid O) = p(H \mid O), \quad q(H \mid G) = p(H \mid G).$$

Presumably these conditions both hold: Before hearing the reader's voice you attributed certain subjective probabilities to Olivier's having married Leigh (high), and to Gielgud's having done so (low). Nothing in what you heard tended to change those judgments: Your judgment about H changed only incidentally to the change in your judgment about O and G. Thus, by (k),

$$q(H) = q(O)p(H \mid O) + q(G)p(H \mid G).$$

$q(H)$ is low because it is a weighted average of $p(H/O)$, which was high, and $p(H/G)$, which was low, with the low value getting the lion's share of the weight: $q(G) = .9$.

2. REPRESENTATION OF BELIEF

In Clark Glymour's book, Bayesianism is identical with personalism, and requires not only updating by conditioning, but also a certain superhuman completeness:

There is a class of sentences that express all hypotheses and all actual or possible evidence of interest; the class is closed under Boolean operations. For each ideally rational agent, there is a function defined on all sentences such that, under the relation of logical equivalence, the function is a probability measure on the collection of equivalence classes. (pp. 68–69)

The thought is that Bayesian personalism must represent one's state of belief at any time by a definite probability measure on some rather rich language. And indeed the two most prominent personalists seem to espouse just that doctrine: De Finetti (1937) was at pains to deny the very meaningfulness of the notion of unknown probabilities, and Savage (1954) presented an axiomatization of preference according to which the agent's beliefs must be represented by a unique probability.

But de Finetti was far from saying that personal probabilities cannot fail to *exist*. [It is a separate question, whether one can be

82

unaware of one's existent partial beliefs. I don't see why not. See Mellor (1980) and Skyrms (1980) for extensive discussions of the matter.] And Savage was far from regarding his 1954 axiomatization as the last word on the matter. In particular, he viewed as a live alternative the system of Bolker (1965) and Jeffrey (1965), in which even a (humanly unattainable) complete preference ranking of the propositions expressible in a rich language normally determines no unique probability function, but rather an infinite set of them. The various members of the set will assign various values throughout intervals of positive length to propositions about which the agent is not indifferent: See Jeffrey (1965, section 6.6) for details.

Surely the Bolker–Jeffrey system is not the last word, either. But it does give one clear version of Bayesianism in which belief states – even superhumanly definite ones – are naturally identified with infinite sets of probability functions, so that degrees of belief in particular propositions will normally be determined only up to an appropriate quantization, i.e., they will be interval-valued (so to speak). Put it in terms of the thesis of *the primacy of practical reason,* i.e., a certain sort of pragmatism, according to which belief states that correspond to identical preference rankings of propositions are in fact one and the same. [I do not insist on that thesis, but I suggest that it is an intelligible one, and a clearly Bayesian one; e.g., it conforms to Frank Ramsey's (1931) dictum (in "Truth and Probability," section 3): "the kind of measurement of belief with which probability is concerned . . . is a measurement of belief *qua* basis of action."] Applied to the Bolker–Jeffrey theory of preference, the thesis of the primacy of practical reason yields the characterization of belief states as sets of probability functions (Jeffrey 1965, section 6.6).

But of course I do not take belief states to be determined by full preference rankings of rich Boolean algebras of propositions, for our actual preference rankings are fragmentary, i.e., they are rankings of various subsets of the full algebras. Then even if my theory were like Savage's in that full rankings of whole algebras always determine unique probability functions, the actual, partial rankings that characterize real people would determine belief states that are infinite sets of probability functions on the full algebras. Here is the sort of thing I have in mind, where higher means better:

$$A, B \quad C$$
$$D$$
$$W \quad W$$
$$-C$$
$$-A, \ -B-D$$
$$(1) \quad (2)$$

This is a miniature model of the situation in which the full Boolean algebra is infinite. Here the full algebra may be thought of as consisting of the propositions A, B, C, D and their truth-functional compounds. W is the necessary proposition, i.e., $W = A \vee - A = C \vee - C$, etc. Here is a case in which the agent is indifferent between A and B, which he prefers to W, which in turn he prefers to $-A$ and to $-B$, between which he is indifferent. But he has no idea where AB, $A \vee - B$, etc., come in this ranking: His preferences about them remain indeterminate. That is what ranking (1) tells us. And ranking (2) gives similar sorts of information about C, D, and their denials: The agent's preferences regarding CD, $C \vee - D$, etc., are also indeterminate. But the two rankings are related only by their common member, W. Thus C and D are preferred to $-A$ and $-B$, but there is no information given about preference between (say) A and C.

That is the sort of thing that can happen. According to (1), we must have $p(A) = p(B)$ for any probability function p in the belief state determined by preferences (1) and (2): See example 3 in chapter 7 of Jeffrey (1965). And according to (2), we must have $p(C) < p(D)$ for any such p: See problem 1 in section 7.7. Then the belief state that corresponds to this mini-ranking (or this pair of connecting mini-rankings) would correspond to the set $\{p : p(A) = p(B)$ and $p(C) < p(D)\}$.

The role of definite probability measures in probability logic as I see it is the same as the role of maximal consistent sets of sentences in deductive logic. Where deductive logic is applied to belief states conceived unprobabilistically as *holdings true* of sets of sentences, maximal consistent sets of sentences play the role of unattainable completions of consistent human belief states. The relevant fact in deductive logic is

Lindenbaum's lemma. *A truth value assignment to a set of sentences is consistent iff consistently extendible to the full set of sentences of the language.*

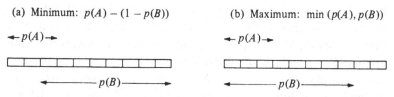

(a) Minimum: $p(A) - (1 - p(B))$

←— $p(A)$ —→

(b) Maximum: $\min(p(A), p(B))$

←— $p(A)$ —→

←——— $p(B)$ ———→

←——— $p(B)$ ———→

Figure 1. $p(AB)$ is the length of overlap between the two segments.

(There is a one-to-one correspondence between consistent truth-value assignments to the full set of sentences of the language and maximal consistent sets of sentences of the language: The truth value assigned is t or f depending on whether the sentence is or is not a member of the maximal consistent set.) The corresponding fact about probability logic is what one might call

De Finetti's lemma. *An assignment of real numbers to a set of sentences is coherent (= immune to Dutch books) iff extendible to a probability function on the full set of sentences of the language.*

(See de Finetti 1972, section 5.9; 1974, section 3.10.)

It is a mistake to suppose that someone who assigns definite probabilities to A and to B (say, .3 and .8, respectively) is thereby committed in Bayesian eyes to some definite probability assignment to the conjunction AB, if de Finetti [1975, (2) on p. 343, and pp. 368–370] is to be counted as a Bayesian. On the other hand, probability logic in the form of the Kolmogorov axioms, say, requires that any assignment to that conjunction lie in the interval from .1 to .3 if the assignments $p(A) = .3$ and $p(B) = .8$ are to be maintained: See Boole (1854, chapter 19), Hailperin (1965), or Figure 1 here. Thus probability logic requires that one or both of the latter assignments be abandoned in case it is discovered that A and B are logically incompatible, since then $p(AB) = 0 < .1$.

Clearly indeterminacies need not arise as that of $p(AB)$ did in the foregoing example, i.e., out of an underlying determinate assignment to the separate components of the conjunction. See Williams (1976) for an extension of de Finetti's lemma to the case where the initial assignment of real numbers $p(S_i) = r_i$ is replaced by a set of conditions of form $r_i \leq p(S_i) \leq s_i$. And note that indeterminacies need not be defined by such inequalities as these. They might equal-

85

ly well be defined by conditions (perhaps inequalities) on the mathematical expectations $E(X_i)$ of random variables – conditions that impose conditions on the underlying probability measures p via the relation $E(X_i) = \int_W X_i \, dp$. More complex special cases arise when "X_i" is replaced by "$(X_i - EX_i)^2$", etc., so that belief states are defined by conditions on the variances etc. of random variables.

Such definitions might be thought of as generalizing the old identification of an all-or-none belief state with the proposition believed. For propositions can be identified with sets of two-valued probability measures. Each such measure, in which the two values must be 0 and 1, can be identified with the possible world in which the true statements are the ones of probability 1. Then a set of such measures works like a set of possible worlds: a proposition. Now Levi and I take belief states to be sets of probability measures, omitting the requirement that they be two-valued. Call such sets "probasitions." The necessary probasition is the set P of all probability measures on the big Boolean algebra in question. P is the *logical space* of probability logic. My current belief state is to be represented by a probasition: a region R in this space. If I now condition upon a proposition E, my belief state changes from R to

$$R/E = {}_{Df}\{P_E : p \in R \text{ and } p(E) \neq 0\}.$$

Perhaps R/E is a proper subset of R, and perhaps it is disjoint from R, but for the most part one would expect the change from R to R/E to represent a new belief state that *partly* overlaps the old one. And in some cases one would expect the operation of conditioning to *shrink* probasitions, e.g., perhaps in the sense that the diameter of R/E is less than that of R when diameter is defined

$$\text{diam}(R) = \sup_{p.q \in R} \| p - q \|$$

and the norm $\| p - q \|$ is defined

$$\| p - q \| = \sup_A | p(A) - q(A) |$$

where "A" ranges over all propositions in the Boolean algebra.

That is how I would ride that hobby-horse. But I would also concede I. J. Good's (1952, 1962) point, that probasitions are only rough characterizations of belief states, in which boundaries are drawn with artificial sharpness, and variations in the acceptability of different members of probasitions and in the unacceptability of various nonmembers go unmarked. In place of probasitions, one might

86

represent belief states by probability measures μ on suitable Boolean algebras of subsets of the space P. Good himself rejects that move because he thinks that μ would then be equivalent to some point μ^* in P, i.e., the point that assigns to each proposition A the definite probability

$$\mu^*(A) = \int_{p \in P} p(A)\ d\mu(p).$$

In our current terminology, the thought is that a definite probability measure μ on P must correspond to a sharp belief state, viz., the probasition $\{\mu^*\}$. To avoid this reduction, Good proposes that μ be replaced by a nondegenerate probasition of type 2, i.e., a nonunit set of probability measures on P; that in principle, anyway, that probasition of type 2 be replaced by a nondegenerate probasition of type 3; and so on. "It may be objected that the higher the type the woolier the probabilities. It will be found, however, that the higher the type the less wooliness matters, provided the calculations do not become too complicated." (Good 1952, p. 114)

But I do not see the need for all that. It strikes me that here, Good is being misled by a false analogy with de Finetti's way of avoiding talk of unknown probabilities (i.e., the easy converse of his representation theorem for symmetric probability functions). De Finetti's point was that where objectivists would speak of (say) coin-tossing as a binominal process with unknown probability of success on each toss, and might allow that their subjective probability distribution for the unknown objective probability x of success is uniform throughout the unit interval, an uncompromising subjectivist can simply have as his belief function the subjectively weighted average of the various putatively objective possibilities, so that, e.g., his subjective probability for heads on the first n tosses would be

$$p(H_1 H_2 \cdots H_n) = \int_0^1 x^n dx = \frac{1}{n+1}$$

In the analogy that Good is drawing, the probasition R is the set of all binomial probability functions p_x where $p_x(H_i) = x$, and μ is the probability measure on P that assigns measure 1 to R and assigns measure $b - a$ to any nonempty subset $\{p_x : a \leq x < b\}$ of R. But whereas for de Finetti the members of R play the role of (to him, unintelligible) hypotheses about what the objective probability function might be, for Good the members of R play the role of hypotheses about what might be satisfactory as a subjective probability function. But if only the members of R are candidates for the role of satisfactory belief function, their subjectively weighted average, i.e., p as above, is not a candidate. [*That p is not binomial*: $p(H_1/H_2) = 2/3 \neq p(H_1)$ $= 1/2$, whereas for each p_x in R, $p_x(H_1/H_2) = p(H_1) = x$.]

The point is that the normalized measure μ over P is not being used as a subjective probability distribution that indicates one's degrees of belief in such propositions as that the true value of x lies between .1 and .3. On the contrary, the uniformity of the μ distribution within R is meant to indicate that one would be indifferent between having to behave in accordance with p_x and having to behave in accordance with p_y for any x and y in the unit interval (where such behavior is determined as well by his utility function); and the fact that $\mu(R) = 1$ is meant to indicate that one would prefer having to behave in accordance with any member of R to having to behave in accordance with any member of $P - R$. (These are rough characterizations because μ assigns measure 0 to each unit subset of P. A precise formulation would have to talk about having to behave in accordance with randomly selected members of intervals, $\{p_x : a \leq x \leq b\}$.)

Then I think Good's apprehension unfounded: I think one can replace probasitional belief states R by probability distributions μ over P that assign most of their mass to R, without thereby committing oneself to a belief state that is in effect a singleton probasition, $\{\mu^*\}$. But this is not to say that one must always have a sharp probability distribution over P: Perhaps Good's probasitions of types 2 and higher are needed in order to do justice to the complexities of our belief states.

On the other hand, I think that in practice, even the relatively simple transition from probasitional belief states to belief states that are sharp probability measures on P is an idle complexity: The probasitional representation suffices, anyway, for the applications of probability logic that are considered in the remainder of this paper.

An important class of such examples is treated in chapter 2 of de Finetti (1937), i.e., applications of what I shall call

de Finetti's law of small numbers. The estimated number of truths among the propositions $A_1, \ldots, A_n$ must equal the sum of their probabilities.

88

That follows from the additivity of the expectation operator and the fact that the probability you attribute to A is always equal to your estimate of the number of truths in the set $\{A\}$: As de Finetti insists, the thing is as trivial as Bayes' theorem. (He scrupulously avoids applying any such grand term as "law" to it.) Dividing both sides of the equation by n, the law of small numbers takes this form:

The estimated relative frequency of truths among the propositions is the average $(p(A_1) + \cdots + p(A_n))/n$ of their probabilities.

Suppose, then, that you regard the A's as equiprobable but have no view about what their common probability is (i.e., you have no definite degree of belief in the A's), and suppose that tomorrow you expect to learn the relative frequency of truths among them, without learning anything that will disturb your sense of their equiprobability. Thus you might represent your belief state tomorrow by the probasition $\{p : p(A_1) = \ldots = p(A_n)\}$, or by a measure on P that assigns a value near 1 to that probasition. But what's the point? If you don't need to do anything on which tomorrow's belief state bears until tomorrow, you may as well wait until you learn the relative frequency of truths among the A's, say, r. At that point, your estimate of the relative frequency of truths will be r (with variance 0), and by mere coherence your degree of belief in each of the A's will also be r. You know all that today.

Note that in the law of small numbers, the A's need not be independent, or exchangeable, or even distinct! The "law" is quite general: as general and as trivial as Bayes's theorem, and as useful.

A mistake that is easy to make about subjectivism is that anything goes, according to that doctrine: Any weird belief function will do, as long as it is coherent.

The corresponding mistake about dress would go like this: Any weird getup will do, if there are no sumptuary laws, or other laws prohibiting inappropriate dress. That's wrong, because in the absence of legislation about the matter, people will generally dress *as they see fit*, i.e., largely in a manner that they think appropriate to the occasion and comfortable for them that occasion. The fact that it is legal to wear chain mail in city buses has not filled them with clanking multitudes.

Then have no fear: The fact that subjectivism does not prohibit people from having two-valued belief functions cannot be expected to produce excessive opinionation in people who are not so inclined, any more than the fact that belief functions of high entropy are equally allowable need be expected to have just the opposite effect. For the most part we make the judgments we make because it would be unthinkable not to. Example: the foregoing application of de Finetti's law of small numbers, which explains to the Bayesian why knowledge of frequencies can have such powerful effects on our belief states.

The other side of the coin is that we generally suspend judgment when it is eminently thinkable to do so. For example, if I expect to learn the frequency tomorrow, and I have no need for probabilistic belief about the A's today, then I am not likely to spend my time on the pointless project of eliciting my current degrees of belief in the A's. The thought is that we humans are not capable of adopting opinions gratuitously, even if we cared to do so: We are generally at pains to come to opinions that strike us as right, or reasonable for us to have under the circumstances. The laws of probability logic are not designed to prevent people from yielding to luscious doxastic temptations – running riot through the truth values. They *are* designed to help us explore the ramifications of various actual and potential states of belief – our own or other people's, now or in the past or the future. And they are meant to provide a Bayesian basis for methodology. Let us now turn to that – focusing especially on the problem ("of old evidence") that Clark Glymour (1980, chapter 3) identifies as a great Bayesian sore point.

3. THE PROBLEM OF NEW EXPLANATION

Probability logic is typically used to reason in terms of partially specified probability measures meant to represent states of opinion that it would be fairly reasonable for people to be in, who have the sort of information we take ourselves to have, i.e., we who are trying to decide how to proceed in some practical or (as here) theoretical inquiry. Reasonableness is assessed by us, the inquirers, so that what none of us is inclined to believe can be summarily ruled out, but wherever there is a real issue between two of us, or whenever one of us is of two minds, both sides are ruled reasonable. Of course, if our opinions are too richly varied, we shall get nowhere;

90

but such radical incommensurability is less common in real inquiry, even in revolutionary times, than romantics would have us think. It is natural to speak of "the unknown" probability measure p that it would be reasonable for us to have. This is just a substitute for more intelligible speech in terms of a variable "p" that ranges over the probasition (dimly specified, no doubt) comprising the probability measures that we count as reasonable. Suppose now that in the light of evidence that has come to our attention, we agree that p should be modified in a certain way: replaced by another probability measure, p'. If $p'(H)$ exceeds $p(H)$ for each allowable value of "p," we regard the evidence as supporting or confirming H, or as positive for H. The degree of support or confirmation is

$$p'(H) - p(H).$$

In the simplest cases, where $p' = p_E$, this amounts to

$$p(H \mid E) - p(H),$$

and in somewhat more complicated cases, where p' comes from p by kinematics relative to the partitioning $\{E_1, \ldots, E_n\}$, it amounts to

$$\Sigma_i p(H \mid E_i) (p'(E_i) - p(E_i)).$$

But what if the evidence is the fresh demonstration that H implies some known fact E? Daniel Garber (1983) shows that – contrary to what one might have thought – it is not out of the question to represent the effect of such evidence in the simplest way, i.e., by conditioning on the proposition $H \vdash E$ that H implies E, so that H's implying E supports H if and only if

(1) $$p(H \mid H \vdash E) > p(H).$$

And as he points out in his footnote 28, this inequality is equivalent to either of the following two (by Bayes' theorem, etc.):

(2) $$p(H \vdash E \mid H) > p(H \vdash E)$$

(3) $$p(H \vdash E \mid H) > p(H \vdash E \mid -H)$$

This equivalence can be put in words as follows:

(I) A hypothesis is supported by its ability to explain facts in its *explanatory domain*, i.e., facts that it was antecedently thought likelier to be able to explain if true than if false.

91

(This idea, suggested by Garber, got more play in an early version of his paper than in the published one.) This makes sense intuitively. Example: Newton saw the tidal phenomena as the sorts of things that ought to be explicable in terms of the hypothesis H of universal gravitation (with his laws of motion and suitable background data) if H was true, but quite probably not if H was false. That is why explanation of those phenomena by H was counted as support for H. On the other hand, a purported theory of acupuncture that implies the true value of the gravitational red shift would be undermined thereby: Its implying *that* is likely testimony to its implying everything, i.e., to its inconsistency.

But something is missing here, namely the supportive effect of belief in E. Nothing in the equivalence of (1) with (2) and (3) depends on the supposition that E is a "known fact," or on the supposition that $p(E)$ is 1, or close to 1. It is such suppositions that make it appropriate to speak of "explanation" of E by H instead of mere implication of E by H. And it is exactly here that the peculiar problem arises, of old knowledge newly explained. As E is common knowledge, its probability for all of us is 1, or close to it, and therefore the probability of H cannot be increased much by conditioning on E before conditioning on $H \vdash E$ (see 4a) – or after (see 4b), unless somehow the information that $H \vdash E$ robs E of its status as "knowledge"

(4a) $\qquad p(H \mid E) \approx p(H)$ if $p(E) \approx 1$

(4b) $\qquad p_{H \vdash E}(H \mid E) \approx p_{H \vdash E}(H)$ if $p_{H \vdash E}(E) \approx 1$

As (4b) is what (4a) becomes when "p" is replaced by "$p_{H \vdash E}$" throughout, we shall have proved (4b) as soon as we have proved (4a) for arbitrary probability functions p. Observe that (4b) comes to the same thing as this:

$$p(H \mid E \,\&\, H \vdash E) \approx p(H \mid H \vdash E) \text{ if}$$
$$p(E \mid H \vdash E) \approx 1.$$

There and in (4), statements of form $x \approx y$ are to be interpreted as saying that x and y differ by less than a ("small") unspecified positive quantity, say, ϵ.

Proof of (4a). The claim is that for all positive ϵ,

$$\text{If } p(-E) < \epsilon \text{ then } -\epsilon < p(H \mid E) - p(H) < \epsilon.$$

To prove the "$< \epsilon$" part of the consequent, observe that

$$p(H \mid E) - p(H) \leq p(H \mid E) - p(HE) \text{ since } p(HE) \leq p(H)$$

$$= \frac{p(HE)}{p(E)} - p(HE) = \frac{p(HE)p(-E)}{p(E)}$$

$$\leq p(-E) \text{ since } p(HE) \leq p(E).$$

Then $p(H/E) - p(H) < \epsilon$. To prove the "$-\epsilon <$" part, note that it is true iff

$$p(H) - \frac{p(HE)}{p(E)} < \epsilon, \text{ i.e., iff}$$

$$p(H - E) + (\frac{p(HE)}{1} - \frac{p(HE)}{p(E)}) < \epsilon, \text{ i.e., iff}$$

$$\frac{p(HE)}{p(E)} (p(E) - 1) < \epsilon - p(H - E)$$

where the left-hand side is 0 or negative since $p(E) \leq 1$, and where the right-hand side is positive since $p(H - E) \leq p(-E) < \epsilon$. Then the "$- \epsilon <$" part of the consequent is also proved.

Yet, in spite of (4), where E reports the facts about the tides that Newton explained, it seems correct to say that his explanation gave them the status of evidence supporting his explanatory hypotheses, H – a status they are not deprived of by the very fact of being antecedently known.

But what does it mean to say that Newton regarded H as the sort of hypothesis that, if true, ought to imply the truth about the tides? I conjecture that Newton thought his theory ought to explain the truth about the tides, *whatever that might be*. I mean that Newton need not have known such facts as these (explained in *The System of the World*) at the time he formulated his theory:

[39.] The tide is greatest in the syzygies of the luminaries and least in their quadratures, and at the third hour after the moon reaches the meridian; outside of the syzygies and quadratures the tide deviates somewhat from that third hour towards the third hour after the solar culmination.

Rather, I suppose he hoped to be able to show that

(T) H implies the true member of $\mathscr{E}$.

where H was his theory (together with auxiliary data) and $\mathscr{E}$ was a set of mutually exclusive propositions, the members of which make various claims about the tides, and one of which is true. I don't mean that he was able to specify $\mathscr{E}$ by writing out sentences that express its various members. Still less do I mean that he was able to identify the true member of $\mathscr{E}$ by way of such a sentence, to begin with. But he knew where to go to find people who could do that to his satisfaction: People who could assure him of such facts as [39.] above, and the others that he explains at the end of his *Principia* and in *The System of the World*. Thus you can believe T (or doubt T, or hope that T, etc.) without having any views about which member of $\mathscr{E}$ is the true one, and, indeed, without being able to give an account of the makeup of $\mathscr{E}$ of the sort you would need in order to start trying to deduce members of $\mathscr{E}$ from H. (Nor do I suppose it was clear, to begin with, what auxiliary hypotheses would be needed as conjuncts of H to make that possible, until the true member of $\mathscr{E}$ was identified.)

David Lewis points out that in these terms, Garber's equivalence between (1) and (2) gives way to this:

(5) $\qquad p(H \mid T) > p(H)$ iff $p(T \mid H) > p(T)$.

Lewis's thought is that someone in the position I take Newton to have been in, i.e., setting out to see whether T is true, is in a position of being pretty sure that

(S) H implies *some* member of $\mathscr{E}$.

without knowing which, and without being sure or pretty sure that (T) *the member of $\mathscr{E}$ that H implies is the true one*. But in exactly these circumstances, one will take truth of T to support H. Here I put it weakly (with "sure" instead of "pretty sure," to make it easy to prove):

(II) If you are sure that H implies *some* member of $\mathscr{E}$, then you take H to be supported by implying the *true* member of $\mathscr{E}$ unless you were already sure it did.

94

Proof. The claim is that

If $p(S) = 1 \neq p(T)$ then $p(H \mid T) > p(H)$.

Now if $p(S) = 1$ then $p(S \mid H) = 1$ and therefore $p(T \mid H) = 1$ since if H is true it cannot imply any falsehoods. Thus, if $1 \neq p(T)$, i.e., if $1 > p(T)$, then $p(T \mid H) > p(T)$, and the claim follows via (5).

Notice one way in which you could be sure that H implies the true members of $\mathscr{E}$: You could have known which member that was, and cooked H up to imply it, e.g., by setting $H = EG$ where E is the true member of $\mathscr{E}$ and G is some hypothesis you hope to make look good by association with a known truth.

Now (II) is fine as far as it goes, but (John Etchemendy points out) it fails to bear on the case in which it comes as a surprise that H implies *anything* about (say) the tides. The requirement in (II) that $p(S)$ be 1 is not then satisfied, but H may still be supported by implying the true member of $\mathscr{E}$. It needn't be, as the acupuncture example shows, but it may be. For example, if Newton had not realized that H ought to imply the truth about the tides, but had stumbled on the fact that $H \vdash E$ where E was in $\mathscr{E}$ and known to be true, then H would have been supported by its ability to explain E.

Etchemendy's idea involves the propositions S, T, and

(F) H implies some false member of $\mathscr{E}$.

Evidently $F = S - T$, so that $-F$ is the material conditional, $-F = -S \vee T$ ("If H implies any member of $\mathscr{E}$ then it implies the true one"), and so the condition $p(F) = 0$ indicates full belief in that conditional. Etchemendy points out that Lewis's conditions in (II) can be weakened to $p(F) \neq 0$ and $p(HS) = p(H)p(S)$; i.e., you are not antecedently sure that H implies nothing false about X (about the tides, say), and you take truth of H to be independent of implying anything about X. Now Calvin Normore points out that Etchemendy's second condition can be weakened by replacing "$=$" by "$\geq$", so that it becomes: Your confidence in H would not be weakened by discovering that it implies something about X. Then the explanation theorem takes the following form:

(III) Unless you are antecedently sure that H implies nothing false about X, you will regard H as supported by implying the truth

95

about X if learning that H implies something about X would not make you more doubtful of H.

The proof uses Garber's principle

(K^*) $p(A \ \& \ A \vdash B) = p(A \ \& \ B \ \& \ A \vdash B)$.

This principle will hold if "$\vdash$" represents (say) truth-functional entailment and if the person whose belief function is p is alive to the validity of *modus ponens;* but it will also hold under other readings of "$\vdash$," as Garber points out. Thus it will also hold if $A \vdash B$ means that $p(A - B) = 0$, on any adequate interpretation of probabilities of probabilities. The proof also uses the following clarifications of the definitions of T and S:

(T) For some E, $E \in \mathscr{C}$ and $H \vdash E$ and E is true.

(S) For some E, $E \in \mathscr{C}$ and $H \vdash E$.

Proof of (III). The claim is this:

If $p(S - T) \neq 0$ and $p(HS) \geq p(H)p(S)$ then $p(HT)$
$$> p(H)p(T).$$

By (K^*), $p(HS) = p(HT)$, so that the second conjunct becomes $p(HT) \geq p(H)p(S)$. With the first conjunct, that implies $p(HT) > p(H)p(T)$ because (since T implies S) $p(S - T) \neq 0$ implies $p(S) > p(T)$.

Note that (III) implies (II), for they have the same conclusion, and the hypotheses of (II) imply those of (III):

(6)   If $p(S) = 1 \neq p(T)$ then $p(S - T) \neq 0$ and $p(HS)$
$$\geq p(H)p(S).$$

Proof. $p(S - T) \neq 0$ follows from $p(S) = 1 \neq p(T)$ since T implies S, and $p(S) = 1$ implies that $p(HS) = p(H) = p(H)p(S)$.

The explanation theorem (III) goes part way toward addressing the original question, "How are we to explain the supportive effect of belief in E, over and above belief in $H \vdash E$, where H is a hypothesis initially thought especially likely to imply E if true?" Here is a way of getting a bit closer:

96

(IV) Unless you are antecedently sure that H implies nothing false about X, you take H to be supported more strongly by implying the truth about X than by simply implying *something* about X.

Proof. The claim is that

If $p(S - T) \neq 0$ then $p(H \mid T) > p(H \mid S)$,

i.e., since T implies S, that

If $p(S) > p(T)$ then $p(HT)p(S) > p(HS)p(T)$,

i.e., by (K^*), that

If $p(S) > p(T)$ then $p(HT)p(S) > p(HT)p(T)$.

But the original question was addressed to belief in a particular member E of $\mathscr{C}$: a particular truth about X, identified (say) by writing out a sentence that expresses it. The remaining gap is easy to close (as David Lewis points out), e.g., as follows.

(7) For any E, if you are sure that E is about X, implied by H, and true, then you are sure that T is true.

Proof. The claim has the form

For any E, if $p(\Phi) = 1$ then $p(\text{for some } E, \Phi) = 1$ where Φ is this:

$E \in \mathscr{C}$ and $H \vdash E$ and E is true.

Now the claim follows from this law of the probability calculus

$p(X) \leq p(Y)$ if X implies Y

in view of the fact that Φ implies its existential generalization.

Here is an application of (III):

Since Newton was not antecedently sure that H implied no falsehoods about the tides, and since its implying anything about the tides would not have made it more doubtful in his eyes, he took it to be supported by implying the truth about the tides.

And here is a corresponding application of (7):

Newton came to believe that H implied the truth about the tides when he came to believe that H implied E, for he already regarded E as a truth about the tides.

97

To couple this with (III), we need not suppose that Newton was antecedently *sure* that H implied something or other about the tides, as in (II). In (III), the condition $p(S) = 1$ is weakened to $p(S) > p(T)$, which is equivalent to $p(S - T) \neq 0$, i.e., to $p(F) \neq 0$.

Observe that in coming to believe T, one also comes to believe S. But if it is appropriate to conditionalize on T in such circumstances, it is not thereby appropriate to conditionalize on S, unless $p(S) = p(T)$, contrary to the hypotheses of (III).

Observe also that although we have been reading "$H \vdash E$" as "H implies E," we could equally well have read it as "$p(E \mid H) = 1$" or as "$p(H - E) = 0$": (K^*) would still hold, and so (III) would still be provable.

4. PROBABILITY LOGIC

Let us focus on the probabilistic counterpart of truth-functional logic. (See Gaifman 1964 and Gaifman and Snir 1982 for the first-order case.)

With de Finetti (1970, 1974) I take expectation to be the basic notion, and I identify propositions with their indicator functions, i.e., instead of taking propositions to be subsets of the set W of all possible "worlds," I take them to be functions that assign the value 1 to worlds where the propositions are true, and 0 where they are false.

Axioms: *The expectation operator is*

linear: $\mathbf{E}(af + bg) = a\mathbf{E}f + b\mathbf{E}g$
positive: $\mathbf{E}f \geq 0$ if $f \geq 0$
normalized: $\mathbf{E1} = 1$

["$f > 0$" means that $f(w) > 0$ for all w in W, and 1 is the constant function that assigns the value 1 to all w in W.]

Definition: *The probability of a proposition A is its expectation, $\mathbf{E}A$, which is also written more familiarly as $p(A)$. De Finetti (1974, section 3.10) proves what he calls "The Fundamental Theorem of Probability":*

98

Given a coherent assignment of probabilities to a finite number of propositions, the probability of any proposition is either determined or can coherently be assigned any value in some closed interval.

(Cf. de Finetti's lemma, in section 2 above.)

A remarkable tight connection between probability and frequency has already been remarked upon. It is provided by the law of small numbers, i.e., in the present notation.

$$\mathbf{E}(A_1 + \cdots + A_n) = p(A_1) + \cdots + p(A_n).$$

That is an immediate consequence of the linearity of $\mathbf{E}$ and the definition of "$p(A_i)$" as another name for $\mathbf{E}A_i$. But what has not yet been remarked is the connection between observed and expected frequencies that the law of small numbers provides.

Example: "Singular Predictive Inference," so to speak. You know that there have been s successes in n past trials that you regard as like each other and the upcoming trial in all relevant respects, but you have no information about which particular trials produced the successes. In this textbook case, you are likely to be of a mind to set $p(A_1) = \cdots = p(A_n) = p(A_{n+1}) = x$, say. As $\mathbf{E}(A_1 + \cdots + A_n)$ $= s$ because you *know* there were s successes, the law of small numbers yields $s = nx$. Thus your degree of belief in success on the next trial will equal the observed relative frequency of successes on the past n trials: $p(A_{n+1}) = s/n$.

In the foregoing example, no particular prior probability function was posited. Rather, what was posited was a condition $p(A_i) = x$ for $i = 1, \ldots, n + 1$, on the posterior probability function p: What was posited was a certain probasition, i.e., the domain of the variable "p." The law of small numbers then showed us that for all p in that domain, $p(A_i) = s/n$ for all $i = 1, \ldots, n + 1$. But of course, p is otherwise undetermined by the condition of the problem, e.g., there is no telling whether the A_i are independent, or exchangeable, etc., relative to p, if all we know is that p belongs to the probasition $\{p : p(A_1) = \cdots = p(A_n) = p(A_{n+1})\}$.

A further example: *Your expectation of the relative frequency of success on the next m trials will equal the observed relative frequency s/n of success on the past n trials in case*

$$(8) \quad p(A_1) = \cdots = p(A_n) = x = p(A_{n+1}) = \cdots = p(A_{n+m}).$$

99

Proof. As we have just seen, the first part of (8) assures us that $x = s/n$, and by the second part of (8), the law of small numbers yields an expected number of successes on the next m trials of $\mathbf{E}(A_{n+1} + \cdots + A_{n+m}) = mx$. Then by linearity of $\mathbf{E}$, the expected relative frequency of success on the next m trials is

$$\mathbf{E}\left(\frac{A_{n+1} + \cdots + A_{n+m}}{m}\right) = \frac{ms/n}{m} = \frac{s}{n},$$

i.e., the observed relative frequency of success in the first n trials.

What if you happen to have noticed which particular s of the first n trials yielded success? Then the first part of (8) will not hold: $p(A_i)$ will be 0 or 1 for each $i = 1, \ldots n$. Still, your judgment *might* be that

$$(9) \qquad \frac{s}{n} = p(A_{n+1}) = \cdots = p(A_{n+m}),$$

in which case the expected relative frequency of success on the next m trials will again be s/n, the observed relative frequency on the first n. But maybe the pattern of successes on the first n trials rules (9) out, e.g., perhaps your observations have been that $p(A_1) = \cdots = p(A_s) = 1$ but $p(A_{s+1}) = \cdots = p(A_n) = 0$, so that you guess there will be no more successes, or that successes will be rarer now, etc. The cases in which (9) will seem reasonable are likely to be ones in which the pattern of successes on the first n trials exhibits no obvious order.

These applications of the law of small numbers are strikingly unBayesian in Clark Glymour's sense of "Bayesian": The results $p(A_{n+1}) = s/n = \mathbf{E}(A_{n+1} + \cdots + A_{n+m})/m$ are not arrived at via conditioning (via "Bayes's theorem"), but by other theorems of the calculus of probabilities and expectations, no less Bayesian in my sense of the term.

The emergence of probability in the mid-seventeenth century was part of a general emergence of concepts and theories that made essential use of (what came to be recognized as) real variables. These theories and concepts were quite alien to ancient thought, in a way in which two-valued logic was not: Witness Stoic logic. And today that sort of mathematical probabilistic thinking remains less homely and natural than realistic reasoning from definite hypotheses ("about the outside world") to conclusions that must hold if the hypotheses do. Perhaps "Bayesian" is a misnomer – perhaps one should simply speak of *probability logic* instead. (Certainly "Bayesian *inference*" is a misnomer from my point of view, no less than from de Finetti's and from Carnap's.) But whatever you call it, it is a matter of thinking in terms of estimates (means, expectations) as well as, or often instead of, the items estimated. Thus one reasons about estimates of truth values, i.e., probabilities, in many situations in which the obvious reasoning, in terms of truth values themselves, is unproductive. The steps from two-valued functions ($= 0$ or 1) to probability functions, and thence to estimates of functions that need not be two-valued, brings with it an absurd increase in range and subtlety. To take full advantage of that scope, I think, one must resist the temptation to suppose that a probasition that is not a unit set must be a blurry representation of a sharp state of belief, i.e., one of the probability measures that make up the probasition: an imprecise measurement (specified only within a certain interval) of some precise psychological state. On the contrary, I take the examples of "prevision" via the law of small numbers to illustrate clearly the benefits of the probasitional point of view, in which we reason in terms of a variable "p" that ranges over a probasition R without imagining that there is an unknown true answer to the question, "Which member of R is p?"

REFERENCES

Bolker, Ethan. 1965. *Functions Resembling Quotients of Measures.* Harvard University Ph.D. dissertation (April).
Boole, George. 1854. *The Laws of Thought.* London: Walton and Maberley. Cambridge: Macmillan. Reprinted Open Court, 1940.
Carnap, Rudolf. 1945. On inductive logic. *Philosophy of Science* 12: 72–97.
de Finetti, Bruno. 1937. La prévision: ses lois logiques, ses sources subjectives. *Annales de l'Institut Henri Poincaré* 7. Translated in Kyburg and Smokler (1980).
de Finetti, Bruno. 1972. *Probability, Induction, and Statistics.* New York: Wiley.
de Finetti, Bruno. 1970. *Teoria delle Probabilità.* Torino: Giulio Einaudi editore s.p.a. Translated: *Theory of Probability.* New York: Wiley, 1974 (vol. 1), 1975 (vol. 2).
Diaconis, Persi, and Zabell, Sandy. 1982. Updating Subjective Probability. *Journal of the American Statistical Association* 77: 822–30.
Gaifman, Haim. 1964. Concerning Measures on First Order Calculi. *Israel Journal of Mathematics* 2: 1–18.

101

Gaifman, Haim, and Snir, Mark. 1982. Probabilities over Rich Languages, Testing, and Randomness. *The Journal of Symbolic Logic* 47: 495–548.

Garber, Daniel. 1983. Old Evidence and Logical Omniscience in Bayesian Confirmation Theory. In *Testing Scientific Theories*, ed. John Earman. Minnesota Studies in the Philosophy of Science series, Vol. X. University of Minnesota Press, 1983.

Glymour, Clark. 1980. *Theory and Evidence*. Princeton: Princeton University Press.

Good, I. J. 1952. Rational decisions. *Journal of the Royal Statistical Assn.*, Series B, 14: 107–114.

Good, I. J. 1962. Subjective Probability as the Measure of a Non-Measurable Set. In *Logic, Methodology and Philosophy of Science*, eds. Ernest Nagel, Patrick Suppes, and Alfred Tarski. Stanford: Stanford University Press. Reprinted in Kyburg and Smokler, 1980, pp. 133–146.

Hacking, Ian. 1967. Slightly More Realistic Personal Probability. *Philosophy of Science* 34: 311–325.

Hailperin, Theodore. 1965. Best Possible Inequalities for the Probability of a Logical Function of Events. *The American Mathematical Monthly* 72: 343–359.

Jeffrey, Richard. 1965. *The Logic of Decision*. New York: McGraw-Hill. University of Chicago Press, 1983.

Jeffrey, Richard. 1968. Probable knowledge. In *The Problem of Inductive Logic*, ed. Imre Lakatos. Amsterdam: North-Holland, 1968, pp. 166–180. Reprinted in Kyburg and Smokler (1980), pp. 225–238.

Jeffrey, Richard. 1970. Dracula Meets Wolfman: Acceptance vs. Partial Belief. In *Induction, Acceptance, and Rational Belief*, ed. Marshall Swain, pp. 157–185. Dordrecht: Reidel.

Jeffrey, Richard. 1975. Carnap's empiricism. In *Induction, Probability, and Confirmation*, ed. Grover Maxwell and Robert M. Anderson, pp. 37–49. Minneapolis: University of Minnesota Press.

Keynes, J. M. 1921. *A Treatise on Probability*. London: Macmillan.

Koopman, B. O. 1940. The Axioms and Algebra of Intuitive Probability. *Annals of Mathematics* 41: 269–292.

Kyburg, Henry E., Jr., and Smokler, Howard E., eds. *Studies in Subjective Probability*, 2nd edition. Huntington, N.Y.: Krieger.

Levi, Isaac. 1974. On Indeterminate Probabilities. *The Journal of Philosophy* 71: 391–418.

Levi, Isaac. 1980. *The Enterprise of Knowledge*, Cambridge, Mass.: MIT Press.

May, Sherry, and Harper, William. 1976. Toward an Optimization Procedure for Applying Minimum Change Principles in Probability Kinematics. In *Foundations of Probability Theory, Statistical Inference, and Statistical Theories of Science*, ed. W. L. Harper and C. A. Hooker, vol. 1. Dordrecht: Reidel.

102

Mellor, D. H., ed. 1980. *Prospects for Pragmatism*. Cambridge: Cambridge University Press.

Newton, Isaac. 1934. *Principia* (vol. 2, including *The System of the World*). Motte/Cajori translation. Berkeley: University of California Press.

Ramsey, Frank. 1931. *The Foundations of Mathematics and Other Logical Essays*. London: Kegan Paul. Also: *Philosophical Papers*, Cambridge University Press, 1990.

Savage, Leonard J. 1954. *The Foundations of Statistics*, New York: Wiley. Dover reprint, 1972.

Skyrms, Brian. 1980. Higher Order Degrees of Belief. In Mellor (1980).

Wald, Abraham. 1950. *Statistical Decision Functions*, New York: Wiley.

Williams, P. M. 1980. Bayesian Conditionalization and the Principle of Minimum Information. *British Journal for the Philosophy of Science* 31.

Williams, P. M. 1976. Indeterminate Probabilities. In *Formal Methods in the Methodology of Empirical Sciences*, Proceedings of the Conference for Formal Methods in the Methodology of Empirical Sciences, Warsaw, June 17–21, 1974, ed. Marian Przetecki, Klemena Szaniawski, and Ryszand Wojcick.

POSTSCRIPT (1991): NEW EXPLANATION
REVISITED

Garber represents explanation itself as evidence that can be conditioned upon, i.e., new logico-mathematical evidence showing that the old empirical evidence is implied by the hypothesis under test. To make that work he complicates the Boolean algebra on which the probabilities are defined. The alternative approach explored here retains the original Boolean algebra but complicates the method of revising probabilities.

Here we see probabilistic judgments as design features of probability distributions and of their evolution – features that need not determine unique distributions or unique courses of evolution. This sort of constructivism makes sense of the possibility that I have a probability ratio in mind, absent the individual terms of the ratio, as when I make a conditional probability judgment without the unconditional judgments suggested by the formula $p(this \mid that) = p(this\ \&\ that)/p(that)$. And it makes sense of the possibility (essay 1) that I have a Bayes factor in mind for a change in odds between *this* and *that* without having definite new or old odds in mind for them. That the value of the ratio or of the Bayes factor is 10, say, is a bit of the

103

design of p or of its evolution that's already in place even if none of the component probabilities are or ever will be.

Here we approach the problem of new explanation by reasoning about "flows" in a "space" of probability distributions over the original Boolean algebra. Points p in such spaces represent momentary judgmental states, and parametrized curves p_t represent histories of developing judgments. The effects of particular observations or passages of reasoning are represented by transformations operating on all points of the space. Where the resulting flow increases the probabilities that points assign a hypothesis, the observation or reasoning that prompted the flow supports the hypothesis.

Example. Why does Einstein's explanation of (*A*) the observed advance in the perihelion of Mercury in terms of (*G*) the general theory of relativity increase confidence in *G?* For vividness, suppose the Boolean algebra is determined simply by the four possibilities for joint truth and falsity of *G* and *A*. A history of observations had imposed a transformation *O* on all distributions p with the result that the transformed distributions Op assigned *A* values near 1 – say, exactly 1,

(1a) $$Op(\text{A}) = 1$$

– without changing the probability of *G* conditionally on *A:*

(1b) $$Op(G \mid A) = p(G \mid A).$$

Otherwise put, *O* is a matter of conditioning on *A:*

(1) $$Op(\text{---}) = p(\text{---}\mid A).$$

Einstein's explanation then imposed a transformation *E* on all distributions p, yielding new distributions Ep assigning enormous (say, infinite) odds on *A* given *G* without changing the odds on *A* given $-G$:

(2a) $$\frac{Ep(A \mid G)}{Ep(-A \mid G)} = \infty,$$

(2b) $$\frac{Ep(A \mid -G)}{Ep(-A \mid -G)} = \frac{p(A \mid -G)}{p(-A \mid -G)}$$

104

Entries in the following tables indicate probabilities assigned to the four possibilities determined by G and A by a succession of distributions, e.g., $p_0(GA) = a$, $p_1(G - A) = 0$.

	A	$-A$
G	a	b
$-G$	c	d

p_0

$\xrightarrow{\ O\ }$

$\dfrac{a}{a+c}$	0
$\dfrac{c}{a+c}$	0

$p_1 = Op_0$

$\xrightarrow{\ E\ }$

	0
	0

$p_2 = Ep_1$

Features (2a) and (2b) don't let us fill the remaining two blanks, but a third feature does:

$$(2c) \qquad \frac{Ep(G\,|\,A)}{Ep(-G\,|\,A)} = \frac{p(G\,|\,A)}{p(-G\,|\,A)} \cdot \frac{1}{p_0(A\,|\,G)}.$$

This tells us to increase the odds on G given A by a factor of $(a + b)/a$. With p as in the middle table, the first factor on the right $= a : c$, so the whole right-hand side $= (a + b) : c$. Dividing each term by the sum of both, we can now fill the two blanks.

a	b
c	d

$\xrightarrow{\ O\ }$

$\dfrac{a}{a+c}$	0
$\dfrac{c}{a+c}$	0

$\xrightarrow{\ E\ }$

$\dfrac{a+b}{a+b+c}$	0
$\dfrac{c}{a+b+c}$	0

Note that the result is the same when we apply the transformations in the opposite order:

a	b
c	d

$\xrightarrow{\ O\ }$

$a+b$	0
c	d

$\xrightarrow{\ E\ }$

$\dfrac{a+b}{a+b+c}$	0
$\dfrac{c}{a+b+c}$	0

Thus the transformations are commutative: $OEp = EOp$.

The successive features of E are decreasingly obvious.

Feature (2a) is compelling: If, in connection with known background conditions, G implies A, then of course the upper right-hand entry in the Ep box must be null, and since the upper left-hand entry isn't – since GA is not dismissed as clearly false – the odds on A given G must be infinite.

Feature (2b) may seem compelling because the denial $-G$ of general relativity theory is no theory and cannot be expected to have consequences relevant to the truth of A; but what's at issue is not the bare logical range of ways in which G might be false but our probabilistic judgment in the matter, which may well assign the bulk of the probability of $-G$ to some other theory which itself may imply A, or imply $-A$. Then (2b) has less scope than (2a).[1]

To see the point of (2c), apply the odds form of Bayes's theorem to p_0:

$$\frac{p_0(G \mid A)}{p_0(-G \mid A)} = \frac{p_0(G)}{p_0(-G)} \cdot \frac{p_0(A \mid G)}{p_0(A \mid -G)}$$

But $p_0(A \mid G)$ was $a/(a + b)$ – a mistake, for as Einstein showed, the correct value is 1. The mistake is forgivable and reparable. To correct it we multiply the right-hand side by $1/p_0(A \mid G)$, increasing the p_0 odds on G given A by a factor of $(a + b)/a$ to get the corrected Ep_0 odds. Anyway, we do that when E is applied before O. When E comes after O, as it did historically, it must be the p_1 odds on G given A that we increase by a factor of $(a + b)/a$ to finally get the correct Ep_1 odds. Commutativity of O and E explains why that works.

Like Garber's, the present account of new explanation is basically qualitative. The distribution p_0, with its odds $a : c$ on the not-yet-enunciated G given the not-yet-noticed A, never existed in anybody's mind; and its notional successor p_1 is long gone, although it may have been entertained – say by Einstein, after envisioning G but before explaining the well established A in terms of it. But the account is offered as methodology, not history – as a rational reconstruction, not an origin myth. We're not doing history-of-science fiction when we consider that none of a, b, c, d ought to be null or

---

1. Thanks to Persi Diaconis for this point.

nearly so, so that the net effect of O followed by E distinctly supports G; nor is it pointless to plug in numbers to see, e.g., that with $a = b = c = d$ the first transformation, O or E, leaves the odds on G even, but the other then improves them to $2:1$.

We knew that conditioning (1) is not always the right way to update, and that conditions (1a, b) are necessary and sufficient for (1) to be the way to go. Now we know that conditions (2a, b, c) are necessary and sufficient for an as yet unnamed updating rule to be the way to go. Call it "reparation" for the moment. The rule is actually determined by those three features; in conjunction with the prior distribution p, features (2a, b) determine the ratios of the entries in the rows of the Ep array, and (2c) determines the ratio of the entries in the first colum. As in the two examples we have worked through, this determines all 4 entries, since they must sum to 1. Clearly the range of application of reparation is much narrower than that of conditioning. But its range seems to coincide in large part with that of the paradox of old evidence, new explanation.[2]

2. For help in getting this far I am indebted to Persi Diaconis, Alan Hájek, Brian Skyrms, and Lyle Zynda, none of whom are to be thought of as therefore endorsing the result.

6

Alias Smith and Jones: The testimony of the senses

Probabilistic accounts of judgment can disappoint dogmatically rooted expectations, i.e., expectations rooted in cases where simple acceptance and rejection are appropriate attitudes toward hypotheses, and where evidence can be expected to simply verify or falsify hypotheses. But probabilistic judgment, freed from the two endpoints, can rate hypotheses anywhere in the unit interval, and can do the same for evidence itself. Where the reliability of signs and witnesses is an issue, dogmatism must in some measure take account of probabilistic considerations; but the risk of paradox is then extreme. Here is such a case, presented as an epistemological parable that can also be read as a parable of scientific hypothesis-testing.

THE PROBLEM[1]

H is the hypothesis that it's hot out. Smith and Jones have each testified as to H's truth or falsity, but you don't regard their testimony as completely reliable. Perhaps you're not even quite sure what either of them said. (You are deep in a tunnel, into which they have shouted their reports.) Let E and F be the propositions that Smith and Jones, respectively, *said* that it's hot. How can you represent your judgment about E and F, and your consequent judgment about H?

CLEAR TESTIMONY

Let's start with the relatively simple case where there's no doubt about what was said. Here it's appropriate to treat the evidence dogmatically, as a matter of learning the truth values of E and F.

First published by R. Jeffrey, in *Erkenntnis*, Vol. 26, pp. 391–399, 1987. Reprinted by permission of Kluwer.Academic Publishers.
1. The Smith and Jones problem was posed by Stewart Cohen.

Dogmatism then fits probabilism as hand in glove: You'll replace your prior judgment $P(H)$ by $P(H \mid D)$, where D is whichever of the conjunctions EF, $E - F$, $F - E$, $-E - F$ is true.

But even in this simplest case, confusion abounds. In considering testimony T about H it's common to confuse the following two.

Final odds: $P(H \mid T)/P(-H \mid T)$
Likelihood ratio: $P(T \mid H)/P(T \mid -H)$.

This may be due to the ambiguity of the colloquial formulation, "The odds on his having spoken truly," as between the final odds and the likelihood ratio.

The odds form of Bayes' theorem identifies the likelihood ratio as the factor by which you multiply your prior odds $P(H)/P(-H)$ in order to get the final odds[2]:

Final odds = likelihood ratio × prior odds.

Thus the likelihood ratio plays the same role in the odds form of the theorem that the relevance quotient $P(T \mid H)/P(T)$ does in the probability form:

Final probability = relevance quotient × prior
probability
$$P(H \mid T) = P(T \mid H)/P(T) \times P(H).$$

If Smith's and Jones's testimonies are relevant to H it must be that when D above is simply E or simply F, the final odds on H are different from the initial odds. By Bayes' theorem this means that neither likelihood ratio, $P(E \mid H)/P(E \mid -H)$ or $P(F \mid H)/P(F \mid -H)$, is 1; it means that $P(E \mid H) \neq P(E \mid -H)$, and similarly for F.

Independence of the two witnesses regarding H means that any dependency between E and F is accounted for by the dependency of each upon H. That's a matter of independence conditionally on H's truth and also on H's falsity:

$$P(EF \mid H) = P(E \mid H)P(F \mid H),$$
$$P(EF \mid -H) = P(E \mid -H)P(F \mid -H).$$

It's a matter of conditional independence. Simple independence in the sense that $P(EF) = P(E)P(F)$ is quite another matter. If Smith

2. The odds form of Bayes' theorem continues to hold when H is being contrasted with any other hypothesis, G: $P(H \mid T)/P(G \mid T) = P(T \mid G)/P(T \mid H) \times P(H)/P(G)$. If $G = -H$, this is the odds form of Bayes' theorem as in the text; but in general, G need not even be incompatible with H.

and Jones are even slightly reliable independent sources of information about the weather, they can't be simply independent.[3]

CONFLICTING CLEAR TESTIMONY

What if Smith and Jones differ: What if Smith says it's hot but Jones says it isn't? It would seem that if they're equally reliable independent witnesses of heat, i.e., if

$$P(E \mid H) = P(F \mid H) = r,$$
$$P(E \mid -H) = P(F \mid -H) = s,$$
$$P(EF \mid H) = r^2, P(EF \mid -H) = s^2,$$

then the conflicting reports should cancel, and the final odds must be the same as the initial ones. Right?

Wrong. By Bayes' theorem the final odds will equal the initial ones iff the likelihood ratio is 1: iff $P(E - F \mid H) = P(E - F \mid -H)$, i.e., $P(E \mid H) - P(EF \mid H) = P(E \mid -H) - P(EF \mid -H)$, i.e., in turn,

$$r - r^2 = s - s^2, \quad r - s = r^2 - s^2, \quad r - s =$$
$$(r + s)(r - s).$$

This last equation holds iff $r = s$ or $r + s = 1$. But $r = s$ means that Smith's statement that it's hot sheds no light on the matter (nor does Jones's). Then in case Smith and Jones are both somewhat reliable witnesses, their contradictory reports will cancel iff $r + s = 1$, i.e., $P(E \mid H) + P(E \mid -H) = 1$, and similarly for F[4]:

When equally reliable independent witnesses contradict each other their testimonies cancel iff probabilities of true and false positive reports sum to 1.

3. *Proof.* Set $P(E \mid H) = e$, $P(E \mid -H) = e'$, $P(F \mid H) = f$, $P(F \mid -H) = f'$, $P(EF \mid H) = ef$, $P(EF \mid -H) = e'f'$, $P(H) = h$. Use the law of compound probability $P(G) = hP(G \mid H) + (1 - h)P(G \mid -H)$ with $G = EF$, $= E$, and $= F$, to get an expression for $P(EF) = P(E)P(F)$ that reduces to $e(f - f') = e'(f - f')$, i.e., an equation that holds iff $e = e'$ or $f = f'$.

4. It's a law of probability that $P(E \mid H) + P(-E \mid H) = 1$, but the condition we're considering is $P(E \mid H) + P(E \mid -H) = 1$, which needn't always hold. Example: Smith and Jones are more reliable when it's hot than when it isn't: $P(E \mid H) = 0.9 = r$ but $P(-E \mid -H) = 0.8 = 1 - s$, so $r + s$ comes to 110%.

110

Now let's consider the case where you aren't sure whether Smith said "rain" or not, or whether Jones said "rain" or not. That means you're not sure whether E is true or false, or whether F is; but presumably you've got some notion, based on the garbled sounds that came booming in from the end of the tunnel. Here's where we can use my (1965, 1983, chapter 11) kinematical scheme:

$$P \xrightarrow{E} Q \xrightarrow{F} R.$$

The sounds from Smith lead you to change your probability of E from $P(E)$ to $Q(E)$ and then the sounds from Jones lead you to change your probability of F from $Q(F)$ to $R(F)$ – where $Q(F)$ is your new probability for F, based on having heard from Smith. If you thought it likely that Smith said "rain" then $Q(F)$ will be higher than $P(F)$.

Let's see in general how the changes from $P(E)$ to $Q(E)$ and from $Q(F)$ to $R(F)$ are propagated over the other propositions in the domain of the probability functions P, Q, R.

For any proposition H in the domain, $Q(H)$ is determined as

$$Q(H) = Q(E)Q(H \mid E) + Q(-E)Q(H \mid -E)$$

by the law of compound probability. If the

Sufficiency conditions[5]

$$Q(H \mid E) = P(H \mid E), \quad Q(H \mid -E) = P(H \mid -E)$$

are met, substitution yields

Generalized conditioning[6]

$$Q(H) = Q(E)P(H \mid E) + Q(-E)P(H \mid -E),$$
$$Q(H)/P(H) = [Q(E)/P(E)]P(E \mid H)$$
$$+ [Q(-E)/P(-E)]P(-E \mid H),$$

5. The sufficiency conditions are so called by Diaconis and Zabell (1982); elsewhere (1970) I've called them "rigidity" conditions. They can be defined for countable partitions generally, i.e., for $\{E_i: i = 1, 2, \ldots\}$ where the E_i are mutually incompatible and jointly exhaustive: $Q(H \mid E_i) = P(H \mid E_i)$. In the case treated in the text the partition has just two members, $\{E_1, E_2\}$, with $E_1 = E$ and $E_2 = -E$.
6. More generally (see note 5), $Q(H) = \Sigma_i Q(E_i)P(H \mid E_i)$ and $Q(H)/P(H) = \Sigma_i e_i P(E_i \mid H)$, where e_i is the ratio $Q(E_i)/P(E_i)$.

which, in turn, imply the sufficiency conditions. Intuitively, the sufficiency conditions say that the entire effect on your probabilities of the sounds from Smith is accounted for by their effect on the propositions E, $-E$: Conditionally on each of those propositions, your new probability functions $Q(\mid E)$ and $Q(\mid -E)$ are the same as the old functions, $P(\mid E)$ and $P(\mid -E)$.

Having changed from P to Q by generalized conditioning on E, you respond to the sounds from Jones by changing from Q to a new probability function, R, by generalized conditioning on F: For any proposition H the sufficiency conditions

$$R(H \mid F) = Q(H \mid F), \ R(H \mid -F) = Q(H \mid -F)$$

are equivalent to generalized conditioning on F,

$$R(H) = R(F)Q(H \mid F) + R(-F)Q(H \mid -F),$$
$$R(H)/Q(H) = [R(F)/Q(F)]Q(F \mid H)$$
$$+ [R(-F)/Q(-F)]Q(-F \mid H).$$

That's how to get from P to R in two steps.[7]

Combining the two steps by laborious but straightforward algebra, we can verify that the ratio of final to initial odds on H can be expressed as a ratio of two other quantities, X and Y[8]:

$X/Y = [R(H)/R((-H)]/[P(H)/P(-H)]$, where
$X = abP(EF \mid H) + aP(E - F \mid H) + bP(-EF \mid H)$
$\quad + P(-E - F \mid H),$
$Y = abP(EF \mid -H) + aP(E - F \mid -H) + bP(-EF \mid -H)$
$\quad + P(- E - F \mid -H),$
$a = [Q(E)/Q(-E)]/[P(R)/P(-E)],$
$b = [R(E)/R(-E)]/[Q(E)/Q(-E)].$

Field (1978) conjectured that a and b above are "input parameters," which represent the experiential inputs derived from hearing Smith and hearing Jones.[9] The thought is that the transitions from

---

7. More generally, sufficiency for the pair $\{Q, R\}$ of an F partition $\{F_j: j = 1, 2, \ldots \}$ means that $R(H \mid F_j) = Q(H \mid F_j)$ for all j. Then $R(H)/Q(H) = \Sigma_j f_j Q(F_j \mid H)$ where $f_j = R(F_j)/Q(F_j)$.

8. Expressed in these same terms, the two separate steps are
$$[Q(H)/Q(-H)]/[P(H)/P(-H)]$$
$$= [aP(E \mid H) + P(-E \mid H)]/[aP(E \mid -H) + P(-E \mid -H)],$$
$$[R(H)/R(-H)]/[Q(H)/Q(-H)]$$
$$= [bQ(F \mid H) + Q(-F \mid H)]/[bQ(F \mid -H) + Q(-F \mid -H)].$$

9. Our a and b are Field's $B^{2\alpha}$ and $B^{2\alpha'}$. (He takes B as the base of the system of natural logarithms.) More generally, relative to an E partition as in note 5, his α_i is $K + \log_B e_i$ where mK is the product $\Pi_i e_i$ of all the e's.

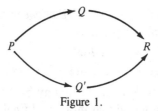

Figure 1.

$P(E)$ to $Q(E)$ and from $Q(F)$ to $R(F)$ are the results of combining those inputs with your ambient judgments when the inputs are received. The ratio of ratios a is meant to dissect out of $P(E)$ and $Q(E)$ the bit that's responsible for that transition, and b is meant to do the same job for the transition from $Q(F)$ to $R(F)$.

Now if b represents just the input derived from the sounds from Jones, we ought to be able to multiply $P(F)/P(-F)$ by it to find what your probability for F would have been after hearing Jones if he had spoken first. Thus X/Y, above, ought to give your final probability for H after hearing Smith and Jones in *either* order. That's Field's (1978) idea. The actual values of a and b were dissected out of the actual sequence: P, Q, R. But, having obtained them so, we ought to be able to use them to find how you would have responded to the same inputs in the other order.

CONFLICTING UNCLEAR TESTIMONY

What's the effect of a seeming contradiction between Smith's and Jones's testimony when the generalized conditioning scheme is used? (See Figure 1.)

Evidently it won't matter in what order they are heard, if changes are made via the parameters a and b, as Field suggests. But there's a question about the effect of the seeming contradiction when the changes are made in *either* order – say, E first, then F.

It's a "seeming" contradiction if the sounds from Smith and from Jones are both unclear. It's not that Smith is heard to say that it's hot and Jones is heard to say that it isn't, i.e., it isn't that $a = 0$ and $b = 1/0$, but that a and b have finite values which are reciprocals: $ab = 1$.

Now suppose (as in the case of conflicting clear testimony) that Smith and Jones are equally reliable independent witnesses:

$$P(E \mid H) = P(F \mid H) = r,$$
$$P(E \mid -H) = P(F \mid -H) = s,$$
$$P(EF \mid H) = r^2, P(EF \mid -H) = s^2.$$

113

Then in the formula for the ratio X/Y between final (R) odds on H and initial (P) odds above we have

$$X = abr^2 + (a + b)r(1 - r) + (1 - r)^2,$$
$$Y = abs^2 + (a + b)s(1 - s) + (1 - s)^2.$$

In the case of conflicting *clear* testimony by equally reliable independent witnesses we found that initial and final odds were equal iff $r + s = 1$. Then let's set $s = 1 - r$, and ask what else must be true in order for initial and final odds to be equal, i.e., for X to equal Y:

$$abr^2 + (a + b)r(1 - r) + (1 - r)^2 = ab(1 - r)^2$$
$$+ (a + b)$$
$$(1 - r)r + r^2.$$

Simplifying, we have $ab(2r - 1) = 2r - 1$, where $2r - 1 \neq 0$ unless $r = s = 1/2$, i.e., unless Smith's and Jones's testimonies are quite worthless. Then we can conclude:

When $r + s = 1$, the condition $ab = 1$ is necessary and sufficient for unclear testimony by two equally reliable independent witnesses to cancel out.

In the case of clear testimony the condition corresponding to $ab = 1$ was that *you hear the witnesses contradict each other*, as when you clearly hear Smith assert that it's hot out and clearly hear Jones deny it. In the case of unclear testimony, when you aren't quite sure what either witness said, the closest you can come to hearing them contradict each other is finding yourself moved as strongly between F and $-F$ by Jones's sounds as you were between E and $-E$ by Smith's, but in the opposite direction. That's when $ab = 1$, i.e., $b = 1/a$. (If in the same direction, $b = a$.)

GARBER'S COUNTEREXAMPLE

Daniel Garber (1980) offers the following counterexample to Field's (1978) conjecture that a and b are input parameters. Suppose it's just Smith out there, shouting his weather report into your tunnel, and suppose that instead of shouting just once, he produces a series of 9 identical shouts: phenomenologically identical irritations of your auditory surfaces. Then you get no more information about the temperature from all 9 irritations than you got from the first. Yet, if

114

a is an input parameter, dependent only on the phenomenology of what you heard, Field's conjecture suggests that you use it 9 times:

$$P \xrightarrow[a]{E} P' \xrightarrow[a]{E} P'' \cdots \xrightarrow[a]{E} P'''''''''.$$

Suppose that $P(E) = 0.3$ and $P'(E) = 0.4$. Then $a = (0.4/0.6)/(0.3/0.7) = 14/9$, and Garber calculates the successive probabilities of E after hearing the same garbled shout again and again. The new odds on E are the old odds times $a = 1.55\ldots$, i.e., starting with odds of $3/7 = 0.43$ (to two decimal places),

$$0.43, 0.67, 1.0, 1.6, 2.5, 3.9, 6.1, 9.4, 15, 23$$

with corresponding probabilities

$$0.30, 0.40, 0.51, 0.62, 0.71, 0.80, 0.86, 0.90, 0.94, 0.96.$$

"That is, after nine repetitions of the same rather uninformative experience, [you] will become virtually certain that [Smith said it's hot]" (Garber, 1980, p. 144).

If we take Garber's point, we'll see the correct sequence of 10 probabilities for E, starting with $P(E) = 0.3$, as

$$0.3, 0.4, 0.4, 0.4, 0.4, 0.4, 0.4, 0.4, 0.4, 0.4.$$

Indeed the value of a that takes you from $P(E)$ to $P'(E)$ is 14/9, but the "input" parameters that take you on from $P'(E)$ to $P''(E)$, etc., are all 1.

Can Field's proposal be saved? The only total rescue I can think of views the phenomenological quality of experience from the perspective of ambient memories, so that the second time you hear Smith's shout it isn't phenomenologically identical with the first, even though your auditory surfaces are identically irritated on the two occasions. But that way out won't interest Field, for he's a physicalist who wants to make irritations of sensory surfaces pull the full weight of the experiential input.

Of course, Field's reparametrization

$$\frac{Q(H)/Q(-H)}{P(H)/P(-H)} = \frac{a\,P(E \mid H) + P(E \mid -H)}{a\,P(E \mid -H) + P(-E \mid -H)}$$

yields the same value of $Q(E)$ as the other formula,

$$Q(H) = Q(E)P(H \mid E) + Q(-E)P(H \mid -E).$$

115

The question is: When is it appropriate to view a and b as input parameters, to which ambient probabilities and phenomenological memories are irrelevant? The Smith and Jones case suggests this answer: *Relative to H, a and b serve as input parameters when E and F are independent conditionally on H and also on* $-H$, i.e., when you think that before they were garbled, the shouts gave independent testimony about H.[10] [Added 1991] But this is no answer to Field's question, about nonrelative input parameters.

REFERENCES

Diaconis, P. and Zabell, S.: 1982, "Updating subjective probability," *Journal of the American Statistical Association* **77**, 822–830.
Field, H.: 1978, "A note on Jeffrey conditionalization," *Philosophy of Science* **45**, 361–367.
Garber, D.: 1980, "Field and Jeffrey conditionalization," *Philosophy of Science* **47**, 142–145.
Jeffrey, R.: 1965, *The Logic of Decision*, New York, 2nd ed., revised University of Chicago Press, Chicago, 1983.
Jeffrey, R.: 1970, "Dracula meets Wolf Man: acceptance vs. partial belief," in M. Swain (ed.), *Induction, Acceptance and Rational Belief*, 157–185, D. Reidel, Dordrecht 1970.

10. This can be generalized to cases where the hypotheses under consideration aren't simply H and $-H$ (see note 2), and where there are more than two of them.

7

Conditioning, kinematics, and exchangeability

PREVIEW

The change ("conditioning") from prior P to posterior $Q = P(\;|\;E)$ is appropriate only if it changes no probabilities conditionally on E. Under similar conditions a generalization of conditioning ("probability kinematics") is appropriate when $Q(E) < 1$. That generalization is pretty nearly equivalent to ordinary conditioning on the extraordinary proposition that $Q(E)$ has a certain value. Whether or not generalized conditioning is sensitive to the order in which successive changes are made depends on how the changes are set, e.g., by probabilities, or by ratios of probabilities. In a finitistic framework, simple and generalized ("partial") exchangeability are characterized and related to probability kinematics.

CONDITIONING[1]

Sometimes an experiment or observation is adequately represented by partitioning the sample space in such a way as to satisfy both of the following conditions.

(1) **Certainty.** *Observing the outcome drives your probability for some cell E of the partitioning to 1.*

(2) **Sufficiency.** *Probabilities conditioned on E remain unchanged.*

First published by R. Jeffrey, in *Causation, Chance, and Credence*, B. Skyrms and W. Harper, eds. Vol. 1, 1988. Reprinted by permission of Kluwer Academic Publishers.
Thanks are due to the National Science Foundation and Princeton University for support of this work, and to Persi Diaconis, David Freedman, David Lewis, Brian Skyrms, Bas van Fraassen, and Sandy Zabell for lots of help. (The conclusion is actually by Diaconis, somewhat rewritten by Jeffrey.)
1. Also known as "conditionalization."

117

Where P and Q are your prior and posterior probability functions, these conditions say that for some cell E and all hypotheses H in the common domain of P and Q, (1) and (2) hold:

(1) $Q(E) = 1$.
(2) $Q(H \mid E) = P(H \mid E)$.
(3) $Q(H) = P(H \mid E)$.

In any such case (3) holds as well, i.e., Q is obtainable from P by conditioning. In fact,

(1) *and* (2) *are jointly equivalent to* (3), *provided* $P(E) > 0$.

Proof. By (1) $Q(H \mid E) = Q(H)$; from this and (2), (3) follows. Conversely, (3) implies $Q(E) = P(E \mid E) = 1$, i.e., (1); and by (3), $Q(H \mid E) = Q(HE)/Q(E) = P(HE \mid E)/P(E \mid E) = P(H \mid E)$, i.e., (2).

The sufficiency condition (2) can be expressed in other forms, which are sometimes more transparent:

(4) *Odds between propositions that imply* E *don't change: If* G *and* H *imply* E *then* $Q(H)/Q(G) = P(H)/P(G)$.
(5) $Q(H)/P(H) = Q(E)/P(E)$ *if* H *implies* E.
(6) $Q(s)/P(s)$ *is constant for all sample points* s *in* E.

Again, it is understood that denominators, e.g., $P(H)$ and $P(E)$ in (5), don't vanish. Of course, the equivalence of (2) with (6) is asserted only where the sample space is countable and P and Q are defined on all of its subsets (sc., *the discrete case*). Summary:

Conditions (2), (4), *and* (5) *are equivalent to each other, and, in the discrete case, to* (6).

Proof. To get (6) from (5) set $H = \{s\}$. Conversely, if $Q(s)/P(s) = k$ for all s in E as in (6) then $Q(E) = \Sigma_{s \in E} kP(s) = kP(E)$ and if H implies E, $Q(H) = \Sigma_{s \in H} kP(s) = kP(H)$, whence (5) follows. Then (5) and (6) are equivalent in the discrete case. Now verify a circle of implications as follows. (2) ⊢ (4): If G and H imply E, then where X is G or H, $Q(X \mid E) = Q(X)/Q(E)$ and $P(X \mid E) = P(X)/P(E)$ or, by

118

(2), $Q(X)/P(X) = Q(E)/P(E)$, whence, (4). (4) $\vdash$ (5): set $G = E$. (5) $\vdash$ (2): Put HE for H in (5) to get $Q(HE)/Q(E) = P(HE)/P(E)$, whence, (2).

DIACHRONIC COHERENCE

Bruno de Finetti's (1937) Dutch book argument for the definition $P(H \mid E) = P(HE)/P(E)$ of conditional probability answers the question of how your acceptance of simple bets is to connect with your synchronic acceptance of conditional bets. David Lewis[2] answers a different question: When should your new probability distribution, Q, be your old distribution, P, conditioned on E?

Both arguments identify your probability for truth of a hypothesis with the price at which you would be willing to buy or sell a ticket that's worth 1 or 0 units of currency depending on whether the hypothesis is true or false. More generally, where the winning ticket is not worth 1 unit, your probability is the fraction p of its winning worth at which you'd buy or sell the ticket; e.g., see ticket 2 below, where the winning worth is p. Your conditional probability for truth of a hypothesis H on condition D is identified with your buying-or-selling price p for a ticket that's worth 1 unit if DH is true, 0 units if $D - H$ is true, and p units if D is false; e.g., see ticket 1 below.

	1			2		3	
	H	$-H$				H	$-H$
D	1	0	D	0	D	1	
$-D$	p		$-D$	p	$-D$	0	

$$=\qquad\qquad +$$

Price:	p	$pP(-D)$	$P(DH)$

In a version that I borrow from Brian Skyrms (1980b), de Finetti's Dutch book argument is the observation that, come what may, ticket 1 is worth exactly the same as tickets 2 and 3 together; and that therefore, your price for 1 ought to be the sum of your prices for 2 and 3: $p = pP(-D) + P(DH)$. Solving for p we find that $p = P(DH)/P(D)$, Q.E.D.

That argument for the relationship $P(H \mid D) = P(DH)/P(D)$ be-

---

2. Lewis's argument is reported in Teller (1976), section 1.3. Freedman and Purves (1969) have a related but different result.

tween your synchronic conditional and unconditional probabilities is less rough than the characterizations of probabilities in terms of money. The argument depends only on those characterizations sometimes holding good – on the assumption that the laws of probability should be no different when those characterizations fail than when they hold.

Underlying Lewis's diachronic argument is the thought that if you have no other use for the money until tomorrow, there is nothing to choose between buying ticket 1 today, when your probability distribution is P, and buying tickets 2 and 3 tomorrow, when your probability distribution is Q – provided you pay the same for 2 and 3 that you would have paid for 1. If today's and tomorrow's prices correspond to P and Q, you'll pay

$$P(H \mid D) \quad P(H \mid D)Q(-D) \quad Q(HD)$$

for the respective tickets. The first of these equals the sum of the other two iff $P(H \mid D) = Q(HD)/Q(D) = Q(H \mid D)$, i.e., the sufficiency condition (2).

Now where the change between P and Q will be due to your learning the truth value of D tomorrow morning, consider two cases. Case 1: D is false. Then $Q(D) = 0$. The three prices are

$$P(H \mid D) \quad P(H \mid D) \quad 0,$$

i.e., exactly what the respective tickets are worth. Then case 1 is vacuous; it places no restrictions on P and Q. Case 2: D is true. Then $Q(D) = 1$, i.e., the certainty condition (1). The three prices are

$$P(H \mid D) \quad 0 \quad Q(H)$$

and the condition under which there's nothing to choose between buying ticket 1 today and buying tickets 2 and 3 tomorrow is that $Q(H) = P(H \mid D)$, Q.E.D.

Note that this conclusion (3) was obtained under assumptions (1) and (2) of certainty and sufficiency. As Teller (1976) points out, the point of the sufficiency requirement is to ensure that in learning that D is true you learn nothing that would change your odds between different ways in which D might come true; see (4).

In the following argument of Lewis's (see Teller 1976) the sufficiency condition plays no explicit role.

Suppose that through an observation or experiment, or in some other way, you expect to learn which cell of a partition is true. You

120

are to announce a complete strategy for changing your probabilities in response to the forthcoming information. Your announcement will specify a new probability distribution for each possibility as to the truth. Suppose that part of what you've announced is that your distribution will be Q if the truth is D.

Lewis. *If P(D) $\neq$ 0 some Dutch strategy will be acceptable to you unless Q(H) = P(H $\mid$ D).*

Proof. Suppose that $Q(H)$ is too big:

$$Q(H) - P(H \mid D) = a > 0.$$

Then the following three tickets, with their accompanying buying or selling instructions, make up a Dutch strategy. [If, instead, $Q(H)$ is too small, so that a is negative, just reverse the buying/selling instructions.]

1	2	3

Worth \$1 or \$0 or $P(H \mid D)$ depending on whether DH or $D - H$ or $-D$ is true. Price: $P(H \mid D)$	Worth \$a if D is true. Price: $aP(D)$	Worth \$1 if H is true. Price: $Q(H)$
SELL NOW	BUY NOW	BUY IF AND WHEN D PROVES TRUE

To verify Dutchness of the strategy, work out your gains and losses from those transactions in the relevant cases, bearing in mind that $P(H \mid D) - Q(H) = -a$:

CASE	GAIN FROM 1	GAIN FROM 2	GAIN FROM 3	NET GAIN
DH	$P(H \mid D) - 1$	$a - aP(D)$	$1 - Q(H)$	$-aP(D)$
$D - H$	$P(H \mid D)$	$a - aP(D)$	$- Q(H)$	$-aP(D)$
$-D$	0	$- aP(D)$	0	$-aP(D)$

Lewis offers a strategy for making one or the other of two books, depending on circumstances: Your book is {Sell 1, Buy 2, Buy 3} or {Sell 1, Buy 2}, depending on whether D is true or false. And since there is no gain or loss from ticket 1 if D is false, the second book has the same value as the singleton {Buy 2}. The sale of 1 and the acquisition of 2 both seem acceptable today, when your probabilities are given by P; and if you acquire 3 tomorrow that will seem fair

121

then, for you will buy 3 only if D is true, hence, only if its price $\$Q(H)$ is acceptable to you then.

Where does the sufficiency condition enter Lewis's argument? It enters with the assumption that today, while your probability distribution is P, you know what tomorrow's Q will be if you learn that D is true. Then your odds between ways in which D might come true are determined by today's judgments: by P. That's why tomorrow's odds between propositions that imply D are the same as today's. That's where the sufficiency condition enters.

SUFFICIENCY AND PAROCHIALISM

When should P be updated to Q by conditioning? The right answer is the sufficiency condition. A different answer, commonly given, is that you should always update by conditioning, i.e., on the strongest proposition E *in the given sample space* to which you now attribute probability 1. That answer is often right, but not always. It's wrong when the E in question violates the sufficiency condition. As an attractive but unreliable rule of probabilistic inference, it's a fallacy. Call it "parochialism."

Clearly, mere certainty isn't enough, e.g., if you learn that E and F are both true, you'll be certain of E, of F, and of their conjunction EF, and yet the conditional probability functions P_E, P_F, and P_{EF} may all be different. If EF is the strongest proposition in the space of which you're now certain, parochialism would have your new probability be P_{EF}. But that may be wrong: Your new probability Q may rightly differ from P_{EF}. The following well-known horror story illustrates how such clashes are sometimes reparable by enriching the sample space.

THE THREE PRISONERS. Two are to be shot and the other freed; none is to know his fate until the morning. Prisoner A asks the warder to confide the name of one other than himself who will be shot, explaining that as there must be at least one, the warder won't be giving away anything relevant to A's own case. The warder agrees, and tells him that B will be shot. This cheers A up a little, by making his judgmental probability for being freed rise from 1/3 to 1/2. But that's silly: A knew already that one of the others would be shot, and (as he told the warder) he's no wiser about his own fate for knowing the name of some other victim.

Diagnosis. A goes wrong by using the 3-point sample space $\{a, b, c\} = \{A$ lives, B lives, C lives$\}$. In that space the warder's statement

122

TABLE 1. *Sample space for the three prisoners problem*

	Warder names B	Warder names C
A lives	$a\bar{b}$ (1/6)	$a\bar{c}$ (1/6)
A dies $\big\{$ B lives		$b\bar{c}$ (1/3)
C lives	$c\bar{b}$ (1/3)	

does eliminate one point, b, and doesn't eliminate either of the others. Following the counsel of parochialism, A conditions on $E = \{a, c\}$. With equal prior weights for the three points, A thus shares the missing 1/3 out equally between a and c. To see why that's wrong we need an ampler sample space, in which one can directly represent propositions about what the warder says; for A's evidence isn't just that B will die, but that the truthful warder names B as one who is to die.

A 4-point space will do if we don't insist on a uniform probability distribution. The points are (say) $a\bar{b}$, $a\bar{c}$, $b\bar{c}$, $c\bar{b}$, where $x\bar{y}$ identifies X as the prisoner who is to go free and Y as the one the warder names. Antecedently, $b\bar{c}$ and $c\bar{b}$ each get weight 1/3, and the weights of $a\bar{b}$ and $a\bar{c}$ sum to 1/3: The way in which that 1/3 is parceled out between them expresses A's opinion about the warder's policy for naming B or C when he can honestly do either.

Presumably A thinks that when the warder can honestly name either B or C as doomed, he's as likely to name one as the other. We'll see that in this case the warder's naming B shouldn't change A's judgment about his own prospects.

The numbers in Table 1 are A's prior probabilities for sample points. A's new probability for living ought to be his old conditional probability given that *the warder names B:*

$$Q(A \text{ lives}) = P(A \text{ lives} \mid \text{warder names } B) = 1/3.$$

That's the same as $P(A \text{ lives})$. A's mistake was to condition on the proposition $E = \{a\bar{b}, c\bar{b}, a\bar{c}\}$ that B will die, i.e., $\{a, c\}$ in the 3-point space, instead of on the stronger proposition $\{a\bar{b}, c\bar{b}\}$ that the warder (truthfully) names B – a proposition that corresponds to no subset of the 3-point space. The trouble with E is that although $Q(E)$ is and ought to be 1 after A hears the warder name B, A's new probabilities $Q(H \mid E)$ conditionally on E won't all agree with the corresponding prior conditional probabilities, $P(H \mid E)$. In either space sufficiency (2) fails for that E – as does the condition of parochialism in the 4-point space. What misleads A is the fact that the condition of parochialism does hold for $E = \{a, c\}$ in the 3-point space.

How do we know that the sufficiency condition fails?

One way is to note that version (4) of that condition fails when E is the proposition $\{a\bar{b}, c\bar{b}, a\bar{c}\}$ that B will die. Hearing the warder name B, not C, A's probability for $a\bar{c}$ drops from $P(a\bar{c}) = 1/6$ to $Q(a\bar{c}) = 0$, while his probability for the warder's naming B climbs from $P(a\bar{b}, c\bar{b}) = 1/2$ to $Q(a\bar{b}, c\bar{b}) = 1$. Then A's odds between propositions $\{a\bar{c}\}$ and $\{a\bar{b}, c\bar{b}\}$ that imply E change from a positive value $(1:3)$ to zero.

Another way is to note that version (5) fails: $Q(H)/P(H)$ vanishes when H is $\{a\bar{c}\}$, but not when H is $\{a\bar{b}, c\bar{b}\}$.

Still another is to note that version (6) fails: $Q(s)/P(s)$ vanishes when s is $a\bar{c}$, but not when s is (say) $a\bar{b}$.

PROBABILITY KINEMATICS

We have seen that natural enlargements of sample spaces can yield propositions satisfying both conditions: sufficiency, and parochialism too. But that doesn't always work.

Probability kinematics is a generalization of conditioning, apt in some cases where observation or experiment or reflection prompts redistribution of probabilities over a countable[3] partition E of a sample space without driving the probability of any cell all the way to 1. As in conditioning, we require that the observation (or whatever) change the judgmental probability distribution over a partition without changing any probabilities conditionally on cells E of the partition:

(7) **Sufficiency.** $Q(H \mid E) = P(H \mid E)$ for each E in $\mathbf{E}$ for which $Q(E) > 0$.

Here we suppose that $P(E)$ is positive for each cell E: There is no possibility of raising $P(E)$ from 0 to a positive value $Q(E)$. But

3. The present restriction to countable partitions is inessential: See Jeffrey (1965, 1983), section 11.8, for informal discussion, and Diaconis and Zabell (1982), section 6, for a rigorous treatment of the general case.

124

certainly $Q(E)$ can be 0 where $P(E)$ is positive. In that case, where $Q(HE)/Q(E)$ is indeterminate, we assign $Q(H \mid E)$ the value $P(H \mid E)$ so that condition (2) will be satisfied.

Sufficiency (7) corresponds to (2). Condition (1) is dropped: Now the observation need not identify the true cell. Sufficiency is equivalent to the following condition.

(8) Kinematics. $Q(H) = \Sigma_E Q(E)P(H \mid E)$.

Proof. As Q is a probability measure, $Q(H) = \Sigma_E Q(E)Q(H \mid E)$. Then (7) implies (8). For the converse, consider any cell D of **E**. By (8) we have $Q(HD) = Q(D)P(HD \mid D)$ since $P(HD \mid E) = 0$ for each cell $E \neq D$, whence $Q(H \mid D) = P(H \mid D)$.

The kinematical formula (8) is the promised generalization of conditioning. Ordinary conditioning is the special case of (8) in which $Q(D) = 1$ for some cell D and so $Q(E) = 0$ for all $E \neq D$.

In the discrete case, (7) is equivalent to[4]

(9) $Q(s)/P(s) = Q(E_s)/P(E_s)$, where E_s is the cell of **E** that s lies in.

Proof. With $H = \{s\}$ and $E = E_s$, (7) immediately yields $Q(s)/Q(E_s)$ $= P(s)/P(E_s)$, i.e., (9). For the converse, note that in the discrete case $Q(HE) = \Sigma_{s \in HE} Q(s)$, so that (9) implies $Q(HE) = \Sigma_{s \in HE} P(s)$ $Q(E_s)/P(E_s) = [Q(E)/P(E)]\Sigma_{s \in HE} P(s) = [Q(E)/P(E)]P(HE)$, whence (7) is immediate.

Our earlier proofs that (2) is equivalent to (4) and to (5) also prove that (7) is equivalent to each of the following.

(10) *For each E in* **E**, *odds between propositions that imply E don't change.*

(11) *For each E in* **E**, $Q(H)/P(H)$ *is constant for all H that imply E, provided* $P(H) > 0$.

In statistical jargon, condition (7) says that the partition **E** is sufficient for the pair $\{P, Q\}$, and that is the same as saying that

4. Diaconis and Zabell (1982), p. 824, (2.2), point out that the connection of (9) with sufficiency is a version of the Fisher–Neyman factorization theorem.

some statistic T that has the cells of $\mathbf{E}$ as its sets of constancy is sufficient for $\{P, Q\}$. Where $\mathbf{E}$ has n cells, the values of T might be the numbers 0 through $n - 1$, or they might be any other n distinct objects, e.g., most simply, the very cells of the partition: $Ts = E$ iff $s \in E \in \mathbf{E}$.

The ratio $Q(s)/P(s)$ in (9) is a useful statistic. Its importance lies in the fact that it is a minimal sufficient statistic for $\{P, Q\}$, in this sense:

Minimality. *Among partitions of a countable space that are sufficient for $\{P, Q\}$ there is a coarsest, i.e, the one whose cells are the sets of constancy of the statistic r, i.e., $r(s) = Q(s)/P(s)$.*

Proof. Let $\mathbf{E}$ be a partition sufficient for $\{P, Q\}$. By (9), r is constant on each cell of $\mathbf{E}$. Thus each of r's sets of constancy is a union of cells of $\mathbf{E}$, and the collection $\mathbf{R}$ of r's sets of constancy is seen to be a partition at least as coarse as any partition that's sufficient for $\{P, Q\}$. That $\mathbf{R}$ is sufficient for $\{P, Q\}$ follows from the equivalence of (7) to (9) and the fact that (9) is satisfied when $\mathbf{E} = \mathbf{R}$.[5]

CONDITIONING ON FUTURE JUDGMENTS

If the sample space is rich enough, probability kinematics is representable as a form of conditioning, i.e., conditioning on the datum D that your new distribution Q will assign values $Q(E_i) = q_i$ to the cells E_i of a partition $\mathbf{E}$.[6]

Here is Brian Skyrms's version of the argument.[7] According to the Reflection Principle [van Fraassen's (1984) term], D will be

5. This proof is from van Fraassen (1980), p. 173. Diaconis and Zabell (1982), p. 824, point out a proof in Blackwell and Girshick (1954), p. 221.
6. If D is to be a subset of the sample space (one to which P assigns a value), Q must be a random variable: Each sample point w must specify a complete hypothesis Q_w about what your new probability function will be. Then $D = \{w: Q_w(E_i) = q_i, i = 1, \ldots, n\}$. An unknown sample point t represents reality (truth), and your actual new probability function will be Q_t. But think of the "Q"s in the text, e.g., in formula (8), as denoting Q_t, your actual new probability distribution, just as "The population of Chicago in 1900" denotes a number, not a function; not a random variable.
7. Skyrms (1980a, appendix 2, and 1980b, pp. 123–7). I. J. Good (1981) makes a similar suggestion (which he told me about in 1963). The earliest such proposal that I know of, by Jaynes (1959) pp. 152–164, "The A_p Distribution: Memory Storage for Old Robots," was independent of probability kinematics.

126

reflected back into your prior distribution P as in (12) below. Suppose, too, that H is independent of D conditionally on each cell of E, as Skyrms's principle (13) stipulates.

(12) Reflection. $P(E_i \mid D) = q_i$ if $P(D) > 0$
(13) Sufficiency. $P(H \mid E_iD) = P(H \mid E_i)$ if $P(E_iD) > 0$

Now if $Q(H) = P(H \mid D)$, the kinematical formula (8) follows from (12) and (13) by straightforward substitution into the law

$$P(H \mid D) = \Sigma_i P(H \mid E_iD)P(E_i \mid D)$$

of the probability calculus. Thus conditioning suffices for the purposes that probability kinematics is meant to serve, if we'll condition on our own future states of judgment conformably with (12) and (13).

As Skyrms points out, condition (13) isn't met for all H, and shouldn't be. In an extreme example H is the hypothesis that $Q(E_i)$ $= q_i$, and it's not D but E_i that's irrelevant to H; for where D implies H, $P(H \mid E_iD) = 1$ but perhaps $P(H \mid E_i) < 1$. And the case is similar whenever it's final (Q) probabilities rather than the corresponding unknown facts that are initially (P) seen as relevant to H, as in Skyrms's example:

[M]y current probability that I will sweat at the moment of arriving at my final probability, conditionally on the fact that Black Bart will not really try to gun me down *and* that that my final probability that he will try to kill me will be 0.999, is *not* equal to my current probability that I will sweat, conditional on the fact that he will not really try to gun me down. The sweating is highly correlated with my final degree of belief rather than the fact of the matter.[8]

But normally there is a natural Boolean algebra A of "objective" hypotheses H for which sufficiency (13) holds.

Diaconis and Zabell (1982) observe that when Q comes from P by conditioning on D, $1/P(D)$ is an upper bound on the ratios of your new to old probabilities on A: $Q(H)/P(H) \leq 1/P(D)$. That's so because $Q(H) = P(H \mid D) = P(HD)/P(D)$, so $Q(H)/P(H) = P(D \mid H)/P(D)$, where $P(D \mid H) \leq 1$. They prove that this condition is necessary and sufficient for Q to be obtainable by extending P beyond A and then cutting back to A by conditioning on a proposition outside A. I'll call that process "superconditioning."

8. Skyrms (1980a), p. 125.

It will make for clarity if we distinguish three Boolean algebras: the algebra **A** of objective hypotheses on which today's and tomorrow's judgmental probability distributions P and Q are defined; the algebra **D** of subjective hypotheses (notably, the hypothesis D about Q) about which you are unlikely to have any judgmental probabilities; and an overarching algebra **X** in which the other two are somehow embedded. The only information we'll need about today's probabilities on **D** is the probability – p, let's say – of D itself. When we condition on D to get the effect of probability kinematics, the exact value of p is of no importance, as long as it is not 0 or 1: not 0, so that we can condition on D, and not 1, so that conditioning on D can make a difference. Instead of "P" we'll use a different designation, "M," for the probability function on **X** that sums up P and p.

If D is of interest only as a bridge over which to pass from P to Q, both of which are defined on **A**, the question arises: Under what conditions on P and Q can **A** be mapped into some algebra **X** on which a probability measure M can be cooked up out of P and Q, which, conditioned on some member X of **X** $-$ **A**, yields on the image of **A** in **X** a distribution that's the image of Q? Answer:

(14) Diaconis and Zabell. *P yields Q by superconditioning (defined below) iff there is an upper bound $b \geqslant 1$ on the ratios $Q(H)/P(H)$ for propositions H in* **A**.

Definition. (All algebras are Boolean σ-algebras.) P yields Q by superconditioning iff there exist: a Boolean σ-isomorphism f from the algebra **A** on which P and Q are defined onto a subalgebra of an algebra **X**: a probability measure M on **X** that corresponds to P on **A**; and an X in **X** that gets Q out of M by conditioning: $M(f(H)) = P(H)$, $M(X) > 0$, $Q(H) = M(f(H) \mid X)$.

Proof of 14. (Throughout, "H" ranges over **A**.) If Q comes from P by superconditioning on X in **X** then

$$Q(H) = M(f(H) \mid X) \leqslant M(f(H))/M(X) = P(H)/M(X),$$

128

i.e., $Q(H)/P(H) \leq b$ with $b = 1/M(X) \geq 1$. For the converse, suppose b satisfies the conditions, with **A** an algebra (i.e., a Boolean σ-algebra) of subsets of W. For $x \neq y$ not in W, define

$$f(H) = H \times \{x, y\} \quad X = W \times \{x\}$$
$$\mathbf{D} = \{f(W), X, f(W) - X, \emptyset\}$$

X = all countable unions of sets of forms $f(H) \cap X$, $f(H) - X$.

So defined, **X** is an algebra of subsets of $f(W)$, and f respects complementation and union: $f(W - H) = f(W) - f(H)$, $f(\cup_i H_i) = \cup_i f(H_i)$. The idea is that each point $\langle w, x \rangle$ or $\langle w, y \rangle$ of $f(W)$ tells the same story that w does, with the extra information that X is true (x) or false (y). Then $f(H) = \{\langle w, z \rangle : w \in H, z = x, y\}$ gives the same information that H does, i.e., $(f(H) \cap X) \cup (f(H) - X)$. Now define

$$M(f(H) \cap X) = Q(H)/b,$$
$$M(f(H) - X) = P(H) - Q(H)/b.$$

The conditions on b ensure that these equations determine M as a probability measure on X. The three conditions at the end of the definition of superconditioning then hold.

QUALMS ABOUT REFLECTION

Goldstein (1983, 1985) and van Fraassen (1984) provide Dutch book arguments for reflection principles. The following argument for (12) is a minor variant of van Fraassen's.

Van Fraassen. If $P(D) \neq 0$, where D specifies new values $Q(E_i) = q_i$ for the cells of a partition $\{E_i : i = 1, \ldots, n\}$, then a Dutch program will be acceptable to you unless $P(E_i \mid D) = q_i$.

Proof. Set $H = E_i$ and $Q(H) = q_i$ in Lewis's book with $a = q_i - P(E_i \mid D) > 0$:

1	2	3
Worth \$1 or \$0 or \$$P(E_i \mid D)$ depending on whether DE_i or $D - E_i$ or $-D$ is true. Price: \$$P(E_i \mid D)$	Worth \$$a$ if D is true. Price: \$$aP(D)$	Worth \$1 if E_i is true. Price: \$$q_i$
SELL NOW	BUY NOW	BUY IF AND WHEN D PROVES TRUE

If D is false (Lewis's third case) bets 1 and 3 are off, and you lose $aP(D)$ on 2. If D is true (Lewis's first two cases) you lose $aP(D)$ in one way or another. Conclusion: The book is Dutch.

But Van Fraassen (1984) calls attention to a troublesome consequence of the reflection principle. To highlight the problem, consider Brian Skyrms's (1987) example:

THE WRATH OF KHAN. If I am now convinced that tomorrow the mindworm of the treacherous Khan will change my probability for H to a certain unreasonable value a, the reflection principle requires me to adopt that unreasonable value already today.

Proof. In (12), set $E_1 = H$. To this D assigns a as its new, Q probability. But by (12) a can't be new after all if $P(D) = 1$, for then $P(H \mid D) = P(H) = q_1 = a$.

The problem isn't that you are conditioning on D. In the present case, as in the problem of the three prisoners, there's a stronger proposition that you're certain of, i.e., here, the Mindworm hypothesis M that both implies and explains D's truth. Although the sufficiency condition fails in the form $Q(\mid D) = P(\mid D)$ perhaps it holds with M in place of D. Perhaps; but that's no help. Even if you condition on M, $P(D)$ will be 1, and that's enough to make the reflection principle imply that $P(E_i) = q_i$.

In "Belief and the will" van Fraassen seemed to suggest that in the light of such examples the reflection principle is to be rejected in case your future judgmental state is thus alien to you. What goes wrong in the Wrath of Khan example is the fact that today's and tomorrow's judgmental states are different in the way in which two different people's states might be different. There is a discontinuity between your points of view today and tomorrow that isn't accounted for by the different values that P and Q assign, but by your present view of the source of that difference. It's not that you can think of no reasonable way in which a might come to be your probability for H tomorrow, but that in fact you think the change will not come about in any such way: The hypothesis R that the transition from P to Q is a reasonable one rules D out.

We can block such applications of the reflection principle by restricting it as follows. [In the 1988 version (15) was garbled. Here it is corrected and further explained in the rest of this paragraph in the light of discussions with Patrick Maher.]

130

(15) Reflection restricted. $P(E_i \mid D) = q_i$ *if* $P(R \mid D) = 1$, *where R is the proposition that the P-to-Q transition specified by D would be reasonable from your present point of view.*

Your criteria for truth of R, for reasonableness of a change to D, must depend on the propositions E_i that D concerns, and on your specific judgments about the particular change. If you are a confident, experienced radiologist who sees D as a possible outcome of the x-ray examination you are beginning, $P(R \mid D)$ will be 1 or near it. In the Wrath of Khan case, $P(R \mid D)$ will be 0 or near it. In general, reasonableness is a genus of specific notions deriving its unity from judgmental consequences encoded in (15). Given generic applicability, certain consequences are expectable; but conditions of applicability are specific, and couched in other terms. It is a familiar pattern. Example: Equiprobability is a generic notion unified by decision-theoretic consequences of its applicability, but its conditions of applicability are specific, depending on the propositions in question and on various physical judgments – e.g., about the center of gravity of some die.

But what about less extreme cases, where you are less than certain of D's truth, or where the Mindworm story is replaced by some more homely hypothesis about fear, fatigue, drugs, alcohol, etc.? Do you get into trouble by merely entertaining one of these hypotheses, i.e., by having $P(D) \neq 0$ when you regard D as pathological?

No. To entertain an analog of the Mindworm hypothesis is just to give some credence to the possibility that you really will be in trouble tomorrow – real trouble, which you don't get into just by entertaining the hypothesis that it may come about. To see that more clearly, let's review the argument for (13), to see where it goes wrong when $P(D \mid R) = 0 < P(D)$. We'll see that where $P(D \mid R) = 0$ your expectation of gain from books that would have 0 expectation under happier circumstances will be negative, and a Dutch book may well have higher expected utility than any available non-Dutch book. Too bad, if so. But it's not thinking that makes it so. Here are the details.

Today, when your probability function is P, the Dutch strategy won't be yours. Today you'll sell 1 and buy 2, but you won't instruct your broker to buy 3 if and when D proves true. You know better. Of course, if you are acting as your own broker tomorrow, and D proves true, you will then regard 3 as a bargain. And you'll

131

then view today's judgment as an aberration from which you have fortunately recovered; you'll buy 3, knowing that you'll thereby suffer the same overall loss of $aP(D)$ that you would have suffered if D had been false. As you see already, that will look sensible to you because your expectation if you do *not* buy 3 would then be the same, i.e., $P(E_i \mid D) - Q(H)$ from the sale of 1, added to a known gain of $a - aP(D)$ from buying 2, a winner.

That's sad; and you're saddened by it already, today, for today you think that if D is true you'll be gulled tomorrow; and you even see today that whether D is true or false, tomorrow you'll think you were gulled today, when you allowed yourself to be guided by the function P in deciding to sell 1 and buy 2 – transactions from which your (P) expectation was 0 but from which you now see that tomorrow's expectation will be $-aP(D)$ on either hypothesis about D.

That's sad, but that's life. If your judgments today and tomorrow are so alien that on each day you regard yourself as irrational on the other, you are in genuine difficulties. Chemotherapy or psychotherapy might help, but on the Mindworm hypothesis and its homely analogs, the laws of probability can't. They can't tell you which (if either) of P, Q is reasonable; nor should you expect them to.

The key is your readiness to identify yourself today with yourself tomorrow. It is only if you do so that the diachronic Dutch strategy in the proof of (12) will be yours, simply. If you don't, today's "you" accepts transactions 1 and 2 but rejects 3 and tomorrow's accepts 3 but rejects the first two; and there's no one agent that you both recognize as accepting the whole book.

SUCCESSIVE UPDATING

What happens when the kinematical formula (8) is applied to two different partitions, $\mathbf{E} = \{E_i: i = 1, \ldots, m\}$, $\mathbf{F} = \{F_j: j = 1, \ldots, n\}$, in succession?

$$P \xrightarrow{\mathbf{E}} Q \xrightarrow{\mathbf{F}} R.$$

It's straightforward to get from P to R in two steps. First get from P to Q via (8) – which, following Hartry Field (1978),[9] we'll find it useful to rewrite as

(17) $Q(H) = P(H)\Sigma_i e_i P(E_i \mid H), \quad e_i = Q(E_i)/P(E_i).$

9. Corresponding to our e_i Field uses $\alpha_i = \log e_i + M$, where M is the constant $(1/m)\Sigma_i \log [q_i/P(E_i)]$.

132

Then get from Q to R by applying the same transformation again, but with $F_j \in \mathbf{F}$ in place of $E_i \in \mathbf{E}$, and with Q and R in place of P and Q:

(18) $R(H) = Q(H)\Sigma_j f_j Q(F_j \mid H), \quad f_j = R(F_j)/Q(F_j).$

To combine these into a single step from P to R, put $Q(F_j H)/Q(H)$ for $Q(F_j \mid H)$ in (18), and then apply (17) with $F_j H$ in place of H:

$$R(H) = \Sigma_j f_j Q(F_j H)$$
$$= \Sigma_j [f_j P(F_j H)\Sigma_i e_i P(E_i \mid F_j H)]$$
$$= \Sigma_j [f_j \Sigma_i e_i P(E_i F_j H)]$$

or, finally,

(19) $R(H) = P(H)\Sigma_{ij} e_i f_j P(E_i F_j \mid H), \quad e_i = Q(E_i)/P(E_i),$
$$f_j = R(F_j)/Q(F_j).$$

This suggests that the effect of the two steps above is the same as that of one big step from P to R via the product $\mathbf{E} \times \mathbf{F} = \{EF : E \in \mathbf{E}, F \in \mathbf{F}\}$ of the two partitions, where in the big step the weight $g_{ij} = R(E_i F_j)/P(E_i F_j)$ of each cell $E_i F_j$ is the product of the weights that its factors have in the two-step transition:

(20) $R(H) = P(H)\Sigma_{ij} g_{ij} P(E_i F_j \mid H),$

$$g_{ij} = R(E_i F_j)/P(E_i F_j) = e_i f_j.$$

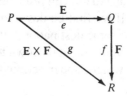

The suggestion is right:

If $\mathbf{E}$ is sufficient for $\{P, Q\}$ and $\mathbf{F}$ is sufficient for $\{Q, R\}$ then $\mathbf{E} \times \mathbf{F}$ is sufficient for $\{P, R\}$; and $g_{ij} = e_i f_j$.

Proof. By definition, $g_{ij} = R(E_i \mid F_j)R(F_j)/P(F_j \mid E_i)P(E_i)$. Now under the two hypotheses, $R(E_i \mid F_j) = Q(E_i \mid F_j) = Q(E_i F_j)/Q(F_j)$ and $P(F_j \mid E_i) = Q(F_j \mid E_i) = Q(E_i F_j)/Q(E_i)$, whence $g_{ij} = Q(E_i)R(F_j)/Q(F_j) P(E_i) = e_i f_j$. Substituting g_{ij} for $e_i f_j$ in (19) we then have (20), according to which $\mathbf{E} \times \mathbf{F}$ is sufficient for $\{P, R\}$.

TABLE 2. *Changes are commutative when specified by ratios*

P	1/2	1/2		Q	19/42	23/42
3/10	1/10	2/10	$e_1 = 5/3$ 1/2		1/6	1/3
7/10	4/10	3/10	$e_2 = 5/7$ 1/2		2/7	3/14
	$f_1 = 21/19$	$f_2 = 36/23$			$f_1 = 21/19$	$f_2 = 36/23$
Q'	21/38	17/38	R		1/2	1/2
11/38	21/190	34/190	$e_1 = 5/3$ 655/874		7/38	13/23
27/38	84/190	51/190	$e_2 = 5/7$ 219/874		6/19	54/161

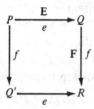

Updating is always commutative when taking a step is a matter of *setting ratios* e_i or f_j of new to old cell probabilities E_i or F_{ij} as in the diagram. Start from P. If we first go down to Q' via $\mathbf{F}$ and then go right via $\mathbf{E}$, we reach the same destination R that we'd have reached by first going right to Q via $\mathbf{E}$ and then down to R via $\mathbf{F}$. That's clear since formula (19) is invariant under permutation of the two steps.

Example. To each point w in the sample space $\{1, 2, 3, 4\}$ the initial distribution P assigns the value $w/10$, as in the upper left of Table 2. The partitions are $E_1 = \{1, 2\}$, $E_2 = \{3, 4\}$ (rows) and $F_1 = \{1, 4\}$, $F_2 = \{2, 3\}$ (columns). Cell probabilities are shown in the margins, e.g., $P(E_2) = 7/10$. Suppose we go right first, making the $\mathbf{E}$ cells equiprobable, and then down, making the $\mathbf{F}$ cells equiprobable. That's a trip from P to Q to R, described in terms of the new cell probabilities: $Q(E_i) = 1/2$, $R(F_j) = 1/2$.

Let's now redescribe the two changes in terms of ratios. The business of making the $\mathbf{E}$ cells equiprobable is a matter of setting $e_1 = 5/3$, $e_2 = 5/7$; and thereafter, making the $\mathbf{F}$ cells equiprobable amounts to setting $f_1 = 21/19$, $f_2 = 36/23$. If instead we had gone from P to Q' by setting the f's just as we did in the second change above we'd have $21/19 = f_1 = Q'(F_1)/P(F_1) = 2Q'(F_1)$, so that $Q'(F_1) = 21/38$ and $Q'(G_2) = 17/38$. We then get from Q' to R if we set $e_1 = 5/3$ and $e_2 = 5/7$ as before. We obtain commutativity by setting ratios of new to old probabilities in the same way, regardless of order. To do so is in effect to set new probabilities – but to set them at values that depend on the order in which the changes are made.

Updating is not generally commutative when each step is a matter of *setting the new probabilities q_i or r_j* of the cells of a partition. The reason is that when we update by first changing probabilities of the **F** cells to r_j and then changing probabilities of the **E** cells to q_i, formulas (17), (18), and (19) give way to

$$S(H) = P(H)\Sigma_j[r_j/P(F_j)]P(F_j \mid H)$$

$$T(H) = S(H)\Sigma_i[q_i/S(E_i)]S(E_i \mid H)$$

$$= P(H)\Sigma_{ij}[q_ir_j/S(E_i)P(F_j)]P(E_iF_j \mid H).$$

Comparing the one-step version of $T(H)$ with (19) we see that while the partition is the same, i.e., $\mathbf{E} \times \mathbf{F}$, the parameter $e_if_j = q_ir_j/P(E_i)$ $Q(F_j)$ has been replaced by a parameter that will be different except when *for all i, j*, $S(E_i)/P(E_i) = Q(F_j)/P(F_j)$.[10]

FROM EPISTEMOLOGY TO STATISTICS

Probability kinematics was first introduced for an in-house philosophical purpose: to show how, in principle, all knowledge might be merely probable, in the face of a priori arguments to the contrary, e.g., those of C. I. Lewis (1946, p. 186), who saw conditioning as the only reasonable way to modify judgmental probabilities by experience. According to Lewis:

The data which themselves support a genuine probability, must themselves be certainties. We do have such absolute certainties, in the sense data initiating belief and in those passages of experience which later may confirm it. But neither such initial data nor such later verifying passages of experience can be phrased in the language of objective statement – because what can be so phrased is never more than probable.

Using probability kinematics, I aimed to show (1957, 1965) how

10. See sections 3 and 4 of Diaconis and Zabell (1982) for details.

135

the familiar language of objective statement needed no supplementation by what C. I. Lewis (loc. cit.) called "the expressive use of language, in which what is signified is a content of experience and what is asserted is the givenness of that content." I saw myself as defending what Carnap had called "physicalism" against phenomenalistic strictures – including, as it turned out, strictures that Carnap himself was to urge against my use of kinematics.[11] The line was what I have lately (1985, p. 114) been calling "radical probabilism," i.e., the epistemological view (essentially, de Finetti's and Ramsey's) that sees judgmental probabilities as our immediate responses to experience, and sees *Protokollsätze* to which we might attribute probability 1 as too thin on the ground to provide a foundation for probable knowledge.

Although it rejects, as misguided, demands for a uniform, nonprobabilistic basis for probability judgments such as relative frequencies are sometimes thought to provide, radical probabilism does undertake to show when and how data about frequencies are reflected in probability judgments – answering the questions "When?" and "How?" in terms of structural features of judgmental probability distributions, whose owners would see statistics as relevant to their opinions. This is the point of de Finetti's work on exchangeability and partial exchangeability over the past 50 years and more – work recently pushed further and generalized by Persi Diaconis and David Freedman in a way that connects it closely to probability kinematics. The remainder of this paper is an elementary introduction to that work, in terms of probability distributions over finitary sample spaces.

A REALISTIC SAMPLE SPACE

The usual sample space for coin-tossing masks the absurdity of von Mises' (1919, 1957) idea that the actual tosses of a coin are simply a readily accessible finite initial segment of an infinite sequence of possible tosses – an infinite sequence that tells how all the untried tosses would have come out, had they been tried. The sample space is the set of all omega-sequences of 0's (tails) and 1's (heads). The

---

11. In correspondence reproduced in Jeffrey (1975).

136

proposition that the first toss is a head is then represented by the set of all infinite binary sequences that start with 1's; and so on.

What would be a realistic sample space for coin-tossing? Where a fixed finite number n of tosses is contemplated, the sample space $\{0, 1\}^n$ consisting of the 2^n distinct sequences of n 0's and 1's is adequate and realistic; but where n is unknown, we need to paste together the infinity of such finite spaces to get a realistic substitute $W = \cup_n \{0, 1\}^n$ for the unrealistic, familiar "Mises" space.

The usual binomial probability distributions B_{np} over the finite sample spaces $\{0, 1\}^n$ (for coin-tossing with probability p for heads) can be pasted together in infinitely many different ways to get distributions over W, i.e., for any nonnegative a's that sum to 1,

$$P(H) = \Sigma_n a_n B_{np}(H).$$

B_{np} is understood to have its usual values on $\{0, 1\}^n$ and to vanish elsewhere in W; the weight a_n is your probability $P(\{0, 1\}^n)$ for the proposition that there are n tosses in all – a proposition that has no counterpart in the Mises space.

Here's a way of looking at it. Each sample space (for coin-tossing, die-rolling, etc.) is the set W of all "words" $w, x, \ldots$, i.e., finite sequences of "letters" from a finite "alphabet" L. If a, b, c, d, e are in L the words $w = \langle a, b, c \rangle$ and $x = \langle d, e \rangle$ will be written as "abc" and "de," and their concatenation $w + x = \langle a, b, c, d, e \rangle$ will be written as "$abcde$."

$A + B$ will be the set of all results of adding a word from B on to the end of a word from A:

$$A + B = \{x + y: x \in A \text{ and } y \in B\}.$$

There is a 0-letter "null" word o, which functions as an identity element:

$$o + w = w + o = w.$$

Powers of sets A of words are defined:

$$A^0 = \{o\}, A^{n+1} = A + A^n.$$

Finally, the realistic sample space W is the result of applying to the alphabet L the operation $n = 0, 1, 2, \ldots,$

$$A^* = \cup_n A^n, \quad W = L^*.$$

137

Example: die-tossing. L is the six-letter alphabet $\{1, 2, 3, 4, 5, 6\}$, representing the possible outcomes of single rolls of a die; L^* represents all possible (finite!) sequences of such outcomes, and subsets of L^* represent various propositions. Thus, $\{1\}^*$ says that if there have been or will be any tosses, they'll all yield aces; $\{2, 3, 4, 5, 6\}^*$ says that the ace never has shown up and never will; and L^n says that the die will have been rolled n times in all. In particular, L^0, i.e., $\{o\}$, is the proposition that the die is never rolled, and L itself is the proposition that the die is rolled just once. $W + \{1\} + W$ says that there is at least one ace (past, present, or future).

In this notation words are interpreted as complete scenarios, covering the past, present, and future, and *nothing marks the present*.[12] Nor does anything mark the present in the more familiar, bizarre scheme in which the actual, finite sequence of outcomes is thought of as an initial segment of an endless sequence, the rest of which shows how untried tosses would have turned out. Note that in the familiar notation there is no way to say how many tosses there are, e.g., our $L + L^*$ (at least 1), or how the last toss turns out, e.g., our $L^* + \{1\}$.

EXCHANGEABILITY[13]

Use the new notation with a finite alphabet L of h letters, which we think of as being given in a definite order. The *tally* of a word w will be a vector

$$(21) \qquad Tw = (N_1w, \ldots, N_hw)$$

where N_iw is the number of times the ith letter of the alphabet occurs in w. The tally function T is thus a vector-valued statistic. Definition: A probability function P makes the letters of the alphabet *exchangeable* iff words with the same tally are always assigned the same probability:

$$(22) \qquad Uniformity \quad P\{x\} = P\{y\} \text{ if } Tx = Ty.$$

---

12. We could mark the present by using ordered pairs of words as scenarios, with the first member representing what has already come to pass and the second representing what still lies ahead. In this notation $\langle w, o \rangle$ is the form of scenarios in which the sequence of tosses has ended, and $\{\langle w + 1, x \rangle: w, x \text{ in } W\}$ is the proposition that there have already been one or more tosses, the most recent of which yielded an ace.

13. This treatment of exchangeability is based on Diaconis (1977). See also Skyrms (1984), ch. 3.

138

Calling P "symmetrical" (Carnap 1950, 1962) is another way of saying that P makes all letters exchangeable.

Of course the detailed form of the tally function doesn't matter; all that matters is the corresponding partition of W, i.e., the function's sets of constancy. The key point is that two words belong to the same cell iff one can be transformed into the other by some rearrangement (permutation) of its letters:

(23) $Tx = Ty$ iff y is an anagram of x.

If letters are movable, as in Scrabble®, the cell to which a word y belongs is determined by what you get by dumping y's letters into an urn. Exhaustively drawing without replacement from that urn, you get the successive letters of some word in that cell; and each word in the cell is obtainable in that way.

The possible values of T form the set $\{Tw: w \text{ in } L^*\}$. To any such value, say t, the distribution P assigns probability $P(T = t) = P(\{w: Tw = t\}) = P(T^{-1}t)$. The equivalence of (7) with (9) in the discrete case implies:

(24) T *is sufficient for any set of symmetrical distributions that assign positive probabilities to the same values of* T.

Note:

(25) *Symmetrical distributions that assign positive probabilities to a value of* T *become identical when conditioned on that value.*

For if P is symmetrical with tally function T, and if $Tw = t$, then by (22), uniformity,

$$P(w \mid T = t) = 1/(\text{the number of words in } T^{-1}t),$$

which is independent of P.

In particular, for $t = (n_1, \ldots, n_h)$ and $n = \Sigma n_i = $ the length of words with tally t, conditioning on $T = t$ reduces any symmetrical distribution P to the *hypergeometric* probability distribution H_t on L^n that characterizes drawing without replacement from an urn which initially contains n balls, of which n_i ($i = 1, \ldots, h$) are labeled with the ith letter of L:

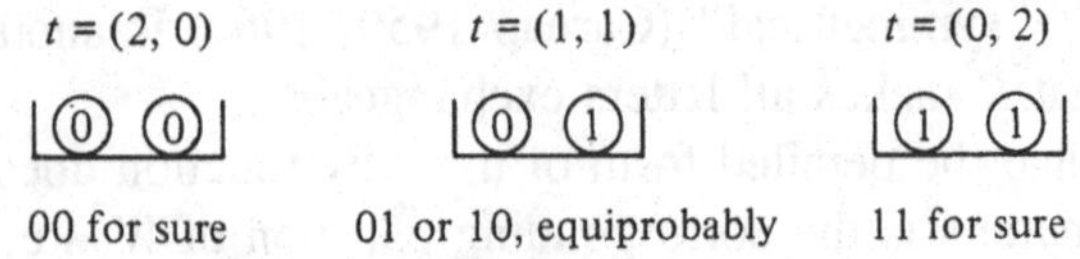

$$t = (2, 0) \qquad\qquad t = (1, 1) \qquad\qquad t = (0, 2)$$

$$\lfloor \textcircled{0} \ \textcircled{0} \rfloor \qquad\qquad \lfloor \textcircled{0} \ \textcircled{1} \rfloor \qquad\qquad \lfloor \textcircled{1} \ \textcircled{1} \rfloor$$

00 for sure   01 or 10, equiprobably   11 for sure

Figure 1.

(26) $P(A \mid T = t) = H_t(A)$
         $= $ (no. of words in $A \cap T^{-1}t$)/
         (no. in $T^{-1}t$).

Furthermore, conditioning on L^n reduces any such P to a definite mixture of hypergeometric distributions H_t (Diaconis 1977):

(27) Finite de Finetti representation.

$$P(A \mid L_n) = \Sigma_t a_t H_t(A), \quad a_t = P(T = t \mid L^n).$$

Example. *Tossing a coin twice (Diaconis 1977).*

$L = \{0, 1\}$ and $t = $ (number of 0's, number of 1's) where 0's are tails and 1's are heads. Any probability distribution over L^2, symmetrical or not, can be identified by the values it assigns to the sample points 00, 01, 10, 11. The symmetrical ones are those for which the middle two values agree: $P\{01\} = P\{10\}$, in which case P(Head first) $= P\{10\} + P\{11\} = P\{01\} + P\{11\} = P$(Head second), $= p$, say. In terms of p, the *binomial* distributions are those symmetrical ones in which $P\{00\} = (1 - p)^2$, $P\{01\} = P\{10\} = p(1 - p)$, $P\{11\} = p^2$. The hypergeometrical distributions form a 3-membered subclass of the symmetrical ones: H_t with $t = (2, 0), (1, 1), (0, 2)$. These correspond to drawing twice without replacement from urns with the compositions shown in Figure 1.

The tetrahedron of Figure 2 represents all the probability distributions over $L^2 = \{00, 01, 10, 11\}$. The shaded triangle represents the symmetrical distributions; its vertices, the hypergeometrical ones; the inscribed parabola, the binomial ones. Each point in the (rigid, weightless) tetrahedron would be the center of mass of a unique system of 4 masses (summing to 1) that might be fixed to its vertices. In these terms theorem (27) says that each distribution $P(\mid L^2)$ in the shaded triangle is the center of mass of some system of masses a_t fixed to its vertices – the masses $a_t = P(T = t \mid L^2)$ being determined uniquely by P. [Thus, the Bayes–Laplace–Johnson–

140

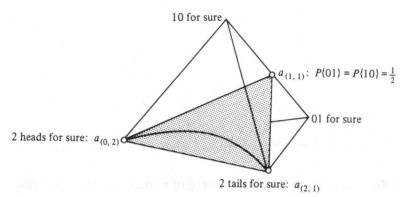

10 for sure

$a_{(1, 1)}$: $P\{01\} = P\{10\} = \frac{1}{2}$

01 for sure

2 heads for sure: $a_{(0, 2)}$

2 tails for sure: $a_{(2, 1)}$

Figure 2.

Carnap distribution, $P(00) = P(11) = 1/3$, $P(01) = P(10) = 1/6$, determines $a_t = 1/3$ for all 3 t's. That's Carnap's (1950, 1962) m^* with 2 individuals and 1 primitive predicate.]

In addition to its unique representation as a mixture of the 3 hypergeometric distributions that are its vertices, each point inside the triangle and under the parabola is obtainable as a mixture of binomial distributions, in infinitely many different ways. That's because each such point lies on various line segments between points on the parabola, and thus is a mixture of each such pair of (binomial) points. Here is a nongeometrical characterization of the symmetrical distributions on L^2 that are mixtures of binomial distributions.

Suppes and Zanotti (1980). *Points in the triangle below, above, and on the parabola represent distributions according to which outcomes of the two trials are positively relevant, negatively relevant, and irrelevant to each other.*

It's $P(\text{Two 1's} \mid L^2) - P(1 \text{ on first toss} \mid L^2)P(1 \text{ on second toss} \mid L^2)$ that determines relevance as positive, negative, or absent.

Corollary. *The mixtures of binomials are the P's that don't make the trials negatively relevant to each other.*

For 3 tosses of a coin the 8-point sample space L^3 can be represented by a 7-dimensional simplex, in which the exchangeable dis-

141

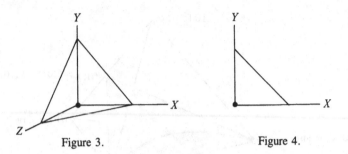

Figure 3. Figure 4.

tributions form a tetrahedron instead of a triangle – through which
the binomial points twist in a curve more complicated than the
parabola of Figure 2 (Diaconis 1977).

De Finetti's representation theorem for arbitrary numbers of
tosses is most often stated in terms of the unrealistic space L^∞ of
omega-sequences of letters:

De Finetti's representation theorem. *Every symmetrical distribution on L^∞ is uniquely representable as a mixture of multinomial
distributions on L^∞.*

In L^∞, talk about a finite sequence w of outcomes is replaced by
talk about the set of all endless prolongations of w. To make that
clear we represent the set of all endless prolongations of w by $\{w\}$ +
L^∞, just as we represent the set of all finite prolongations of w by $\{w\}$
+ L^*. Thus in (28) and (29) below, $P(\{w\} + L^\infty) = P$(the infinite
word begins with w) = $P(X_1 = w_1, \ldots, X_n = w_n)$ in the usual
notation. For vividness we state the theorem in two special cases: L
= $\{0, 1\}$ and $L = \{0, 1, 2\}$.

(28) De Finetti's theorem, infinite case, $L = \{0, 1, 2\}$. *For each
symmetrical distribution P on L^∞ there is a unique probability distribution M over the simplex $S = \{(x, y, z): x + y + z = 1, x, y, z \geqslant 0\}$
of Figure 3 such that for any word w in L^* with tally $Tw = (a, b, c)$,*

$$P(\{w\} + L^\infty) = \int_S x^a y^b z^c \mathrm{d}M.$$

In Figure 3, S is the equilateral triangle with vertices 1 unit from the
origin along the three axes. Where $L = \{0, 1\}$ the corresponding
simplex is the line segment with ends 1 unit from the origin along
the two axes of Figure 4.

142

(29) De Finetti's theorem, infinite case, $L = \{0, 1\}$. *For each symmetrical P on L^∞ there is a unique probability distribution M over the simplex $S = \{(x, y): x + y = 1, x, y, \geq 0\}$ of Figure 4 such that for any word w in L^* with tally $Tw = (a, b)$,*

$$P(\{w\} + L^\infty) = \int_S x^a y^b dM.$$

For the case of a 3-letter alphabet, (28) says that any symmetrical P on L^∞ assigns to $\{w\} + L^\infty$ an M-weighted average of the values $x^a y^b z^c$ that would be assigned to it by the i.i.d. distributions for which the probabilities of the three possible outcomes of a single trial are x, y, z. Thus, any symmetrical P is realizable as the distribution for endless drawings with replacement from an urn containing balls labeled with letters of the alphabet, where the ratio $x:y:z$ of numbers of balls with the different labels was determined randomly in accordance with the distribution M.

For symmetrical P the probability $P(\{w\} + L^\infty)$ as in (28) or (29) serves as an approximation to the corresponding probability in the realistic sample space L^* [14]:

(31) Finite form of de Finetti's theorem. [15] *If P is symmetrical, if the length m of w is less than k, and if $P(\{w + x\}) = 0$ for no word $w + x$ of length k, then (28) or (29) approximates $P(\{w\} + L^* \mid L^k)$ with error less than $2m/k$ times the size of the alphabet.*

The exact value of $P(\{w\} + L^* \mid L^k)$ is an M-mixture of the values assigned to $\{w\} + L^{k-m}$ by the hypergeometric distributions H_t where in the 3-letter case $t = (x, y, z)$ with $x + y + z = k$. These H_t are the distributions appropriate for drawing without replacement from urns initially containing k balls marked with the respective letters of the alphabet in the ratios $x:y:z$. $P(\mid L^k)$ is realizable as the distribution for drawing without replacement from such an urn, for which the ratio $x:y:z$ was thought to be determined randomly in accordance with the distribution M. If $w + x$ is much longer than w (if $k \gg m$) then drawing without replacement won't be much differ-

---

14. Here "P" denotes distinct probability distributions over two different spaces, L^∞ and L^*, in which the proposition that the sequence of outcomes begins with w is represented by the sets of $\{w\} + L^\infty$ and $\{w\} + L^*$, respectively.
15. Diaconis (1977) and Diaconis and Freedman (1980a) prove this theorem and derive the infinite form as a corollary, and point out other proofs of finite forms of the theorem by de Finetti (1972, p. 213) and others.

143

Figure 5.

ent from drawing with replacement (Freedman [1977]), and the expression (28) for $P(\{w\} + L^\infty)$ will closely approximate $P(\{w\} + L^* \mid L^k)$.[16]

MARKOV EXCHANGEABILITY

The foregoing treatment of exchangeability can be generalized by keeping (22) but dropping (21) – allowing T to be other statistics on L^* that have finite-dimensional vectors as values, provided $Tx = Ty$ is a *congruence relation*, i.e., provided

(31) $T(x + w) = T(y + w)$ if $Tx = Ty$

for all w, x, y in L^*.[17] Assumption (31) is automatically satisfied in the special case (22) of complete exchangeability. Here we consider a more general case of that sort studied by de Finetti (1938, 1972), Freedman (1962), Diaconis and Freedman (1980a, b), and Zaman (1984).

Example (Diaconis and Freedman 1980, Diaconis and Zabell 1986): *Tack-Flicking.* A thumbtack is flicked repeatedly – always from the position it landed in after the previous trial, i.e., point down (0) or up (1): See Figure 5. Here a 2-letter alphabet $L = \{0, 1\}$ is apt, with w in $L^* - \{o\}$ reporting the initial position of the tack (first letter) and the outcome of the nth trial ($n + $ 1st letter). If you think the first outcome may depend on whether the tack starts with point to the floor or not, and think that each outcome is influenced by earlier ones only through its immediate predecessor, your tally function might assign to w in L^* a vector with entries indicating its

16. See Freedman (1977): "When drawing a sample of 1000 from a population of 100 000 000, there is almost no difference between drawing with or without replacement [i.e., the error is between 0.00498 and 0.00500]. When drawing a sample of 5000, there is a substantial difference in variation norm [i.e., the error is between 0.117 and 0.125]."

17. Condition (31) here is a translation into the L^* idiom of Diaconis and Freedman's (1982, p. 207) condition (1.2). It is needed in order to prove the representation theorem in the infinite case.

144

Figure 6.

first letter (i) and the numbers of transitions from down to down (00), down to up (01), etc.:

Markov tally function

$$Tw = t = (i; \#00, \#01, \#10, \#11).$$

Thus $Tw = (0; 1, 2, 1, 0)$ for $w = 00101$ or $w = 01001$, and $Tw = (1; 0, 1, 2, 1)$ for $w = 11010$ or $w = 10110$. In general, where $Tw = (i; a, b, c, d)$, the length of w will be $a + b + c + d + 1$. Then words of the same tally must have the same length. One can show that if $Tx = Ty$ then x and y must end with the same letter. It follows that Markov tally functions T satisfy the congruence condition, (30).

For Markov tally functions T over $\{0, 1\}^*$, Diaconis and Freedman (1980, pp. 239–240) contrive urn models that make the words in $T^{-1}t$ equiprobable, as follows. Balls labeled 0 or 1 are drawn without replacement from two urns, themselves labeled 0 and 1. Where i is the first component of t, the first draw is from urn i. Each later draw is from the urn bearing the same label as the ball just previously drawn. The initial contents of the two urns are determined by t as in Figure 6. You are to keep drawing until instructed to draw from an empty urn. If the other urn is empty, too, *the letter i, followed by the labels on the successive balls, spells out a word in* $T^{-1}t$. That's success. If the other urn isn't empty then, you've failed to generate a word in $T^{-1}t$. Then abort that attempt, and try again.

Example. For $t = (0; 1, 2, 1, 0)$ the two urns, and the graph, are composed as in Figure 7. (For the present, ignore the suggestion that one ball is stuck to the bottom of urn 0.) Since $i = 0$, w begins with a 0 and we draw its second letter from urn 0.

Case 1. The 0 is drawn, yielding 00 so far. Then the next draw is from urn 0, too, where only 1's remain: That's 001 so far. The next draw takes the 0 from urn 1 (0010, so far) and the last takes the remaining 1 from urn 0, leaving both urns empty. Success. Result: $w = 00101$.

145

Urn 0: |⓪①①| Urn 1: |⓪|

Figure 7.

Case 2. A 1 is drawn: 01. Then the next draw empties urn 1 (010), and the next takes the 0 or the 1 from urn 0. There are two possibilities.

Case 2a. The 0 is taken: 0100. Then the last draw empties urn 0. Success. Result: $w = 01001$.

Case 2b. The 1 is taken: 0101. Then the next draw is to be from urn 1, which is empty – while a 0 remains in urn 1.

There are only two successful cases: 1 and 2a. For long enough runs of this process, your prior judgmental probability that those two cases will occur in a ratio between (say) 999:1000 and 1000:999 will be as close as you like to 1.

That was a crude version of the urn model. In fact, Diaconis and Freedman (1980b, p. 240) arrange matters so that you can't fail: You'll never be asked to draw from an empty urn. [The following vivid account of that modification was suggested by Diaconis. Zaman (1984), section 7, extends the modification to the general case, of any finite number of letters.]

The Glued Ball Method (where j is the letter that isn't i). If $\#ij = \#ji$ in t, glue one of the i balls to the bottom of urn j; if $\#ij = 1 + \#ji$, glue one of the j balls to the bottom of urn i. Glued balls don't come out. If you are ever instructed to draw from an urn containing only a glued ball, use that ball's label as the word's last letter.

In the example, i was 0 and $\#ij$ was 2 while $\#ji$ was 1, so we glued a j ball to the bottom of urn i. In cases where no gluing is called for, no sequence of draws can fail; but where gluing is called for there are ways to fail without it, e.g., case 2b above, which cannot arise when a ball is glued as in Figure 7. (With 010 as first three letters, we must draw from urn 0, where, since the "1" ball is glued, we must draw the 0, which puts us in the successful case, 2a.)

The process can equally well be described in graphical terms: See Figure 8, where the letters of w are to be the labels of the nodes you visit, starting at node i and always following arrows.

146

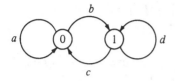

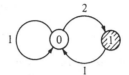

Figure 8.   Figure 9.

The Shaded Node Method (where $j \neq i$). Shade node i if $\#ij = \#ji$; shade node j if $\#ij = 1 + \#ji$. When you traverse an arrow, reduce its numerical label by 1; and erase arrows labeled 0. Never traverse an arrow labeled 1 from an unshaded node to a shaded one when you have another option. Success is ending with no arrows; failure is finding that there's one you can't traverse from where you've ended.

Example. $t = (0; 1, 2, 1, 0)$ again. Then the relevant graph is Figure 9, where node 1 is shaded for the same reason that a "1" ball was stuck to the bottom of urn 0 in Figure 7. Note that in cases 1 and 2a above, all arrows would eventually vanish even without the shading; and note that case 2b cannot now arise, for you may not enter the shaded node while you have the option of generating another 0 instead.

Let M_t be the probability distribution over L^* that corresponds to the urn or graph model described above, i.e., uniform on $T^{-1}t$ and 0 elsewhere in L^*. *Markov symmetry* or "*partial*" *exchangeability relative to a Markov tally function T* is a matter of uniformity (22) relative to T. Corresponding to (26) and (27) in the case of simple exchangeability we have

(32) $P(\mid T^{-1}t) = M_t$ for any Markov-symmetrical P.

(33) De Finetti representation for Markov-symmetrical P.

$P(\mid L^n) = \Sigma_t a_t M_t$ where $a_t = P(T = t \mid L^n)$ and t ranges over the tallies of words of length n.

Corresponding to (29) we have (Diaconis and Freedman 1980b):

(34) De Finetti representation, infinite case: *If P is Markov-symmetrical on L^∞ and P(0 infinitely often) = 1, there is a distribution M on $X = [0, 1]^2$ relative to which P is a mixture of 2-state Markov chains:*

147

$$P(\{w\} + L^\infty \mid \{i\} + L^\infty) =$$
$$\int_X p^a(1 - p)^b q^c(1 - q)^d \mathrm{d}M$$

for all w in L*, where (i; a, b, c, d) = Tw, and where P and q are the transition probabilities of Figure 10.

And corresponding to (30) we have the integral in (34) as an approximation to $P(\{w\} + L^* \mid \{i\} + L^{k-1})$ when k is much greater than the length of w.

PARTIAL EXCHANGEABILITY

Partial exchangeability is a genus, i.e., exchangeability relative to one or another tally function; simple and Markov exchangeability are two of the oldest and best-studied species in that genus. Others are described in Diaconis and Freedman's (1980b) excellent introduction to the whole topic, on which the foregoing account of Markov exchangeability is based, and in a later paper (1982), where they describe the genus in dazzling generality, survey the known species, and report on the state of the art.

The art is a matter of (1) finding de Finetti representations for different species of partial exchangeability, or of (2) going the other way around. In the case of simple exchangeability, task (1) starts with the characterization

(35) $P\{x\} = P\{y\}$ if x is an anagram of y

and ends when the hypergeometric and multinomial distributions are identified as the extreme points in L^* and L^∞, various mixtures of which yield the various symmetrical distributions. But the thing may really have started in the opposite direction, (2), with de Finetti wondering how to characterize mixtures P of multinomial distributions in L^∞ directly, discovering that they are exactly the distribu-

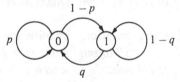

Figure 10.

148

tions satisfying (35), and only then wondering (1) about the extreme points in L^*, the different mixtures of which are the different P's satisfying (35). And as Zaman [1984] suggests, (2) is the much more likely direction in the case of Markov exchangeability.

Anyway, de Finetti representations give special interest to concepts of partial exchangeability, and, so, to tally functions. As (35) suggests, concepts of partial exchangeability can also be characterized by the groups of transformations under which the distributions in question are invariant – groups that define congruence relations on the sample space. (Here von Plato's contribution to the present volume is relevant.)

Even when a species of partial exchangeability is characterized by a tally function T, a further generalization is easy: The uniformity condition (22), "$P\{x\} = P\{y\}$ if $Tx = Ty$," can be dropped in favor of the weaker sufficiency condition (7), which it implies – in the following form (where $T^{-1}t$ is the proposition that T assumes the value t).

(36) For each t, conditioning on $T^{-1}t$ (the hypothesis that $T = t$) makes all distributions in the species identical.

Example. Mixtures of Poisson distributions (Diaconis and Freedman 1980, pp. 246–7). But the congruence condition is indispensable for the representation theorems. That is condition (31) here, or condition (1.2) in Diaconis and Freedman (1982), p. 207.

Sufficiency = partial exchangeability = applicability of probability kinematics. Then where we have a representation theorem for a species of partial exchangeability, probability kinematics can have a special sort of rationale.[18]

Example: The die may be loaded. Perhaps you think that someone who had all relevant information about the die, and had enough time and knowledge of physics to use it, would have a multinomial judgmental probability distribution on $L^* = \{1, 2, 3, 4, 5, 6\}^*$ in which the six possible outcomes of each toss need not be equiprobable. Under such circumstances you may well seek your own judgmental probability distribution among the mixtures of multinomial

---

18. For various examples see Diaconis and Zabell (1986).

distributions. By de Finetti's representation theorem, the statistics (21) $T = (N_1, \ldots, N_6)$ is then sufficient for the set of all candidates for the position of your judgmental probability distribution P over L^*. Note that a definite P may be more than you need, e.g., you may be content with the conditional distributions $P(\mid L^n)$.

CONCLUSION

Partial exchangeability and probability kinematics are a single thing seen from different practical perspectives.

In each case there is a statistic (tally function) T for which the conditional probabilities

$$P(\mid T^{-1}t)$$

play essentially the same, pivotal role. In kinematics these conditional probabilities remain valid for the new probability distribution. Thus the difference between the old and new distributions is determined by the probabilities they assign to T's sets of constancy. In the case of partial exchangeability these conditional probabilities are the extreme points, and every probability distribution that's partially exchangeable relative to T is a convex combination of extreme points.

There is no magic machine to produce useful statistics T, but there is a collection of examples that have proved useful here and there, e.g., familiar classical probability models and their sufficient statistics. All are grist for the mill, i.e., kinematics or T-exchangeability, depending on the point of view.

Consider the kinematical interpretation of the example discussed under Markov exchangeability above of a thumbtack that is flicked repeatedly, always from the position in which it landed. Many statisticians have treated this as a case of simple exchangeability. Reading the present analysis might give them pause: Things *aren't* exchangeable if the last toss matters. If that's all that matters then the tally T of the number of transitions (0 to 0, 0 to 1, 1 to 0, 1 to 1) is a sufficient statistic. Here the uniform conditional probability $P(\mid T^{-1}t)$ specifies P except for four numbers, which must be thought about and specified directly. The kinematical change comes through introspection and insight.

There may be other partial exchangeabilities that are worth recy-

150

cling as kinematics (e.g., among those surveyed in Diaconis and Freedman, 1982).

Probability kinematics and partial exchangeability have different ends in view. Kinematics aims at sensible changes in belief, making full use of relevant past experience. Partial exchangeability is a subjectivistic version of model building.

Both can draw on the same rich lore of applied statistics, accumulated over the years, for ideas about useful statistics T.

REFERENCES

Achinstein, Peter and Owen Hannaway (eds.) (1985) *Observation, Experiment, and Hypothesis in Modern Physical Science* (Cambridge, Mass.: Bradford-MIT Press).

Bernardo, J. M., M. H. DeGroot, D. V. Lindley, and A. F. M. Smith (eds.) (1985) *Bayesian Statistics 2* (Amsterdam: Elsevier, North-Holland).

Blackwell, David and M. A. Girshick (1954) *Theory of Games and Statistical Decisions* (New York: Wiley).

Carnap, Rudolf (1950) *Logical Foundations of Probability* (Chicago: University of Chicago Press; 2nd ed., 1962).

de Finetti, Bruno (1937) "La prévision: ses lois logiques, ses sources subjectives," *Annales de l'Institut Henri Poincaré* **7**, 1–38. Translated in Kyburg and Smokler (1964, 1980).

de Finetti, Bruno (1972) *Probability, Induction and Statistics* (New York: Wiley).

de Finetti, Bruno (1938) "Sur la condition d'équivalence partielle," *Actualités Scientifiques et Industrielles,* No. 739 (Paris: Hermann & Cie.) Translated in Jeffrey (1980).

Diaconis, Persi (1977) "Finite forms of de Finetti's theorem on exchangeability," *Synthese* **36**, 271–281.

Diaconis, Persi and David Freedman (1980a) "De Finetti's generalizations of exchangeability," Jeffrey (1980), 233–249.

Diaconis, Persi and David Freedman (1980b) "Finite exchangeable sequences," *Annals of Prob.* **8**, 745–764.

Diaconis, Persi and David Freedman (1982c) "Partial exchangeability and sufficiency," Proceedings of the Indian Statistical Institute Golden Jubilee International Conference on *Statistics: Applications and New Directions,* Calcutta, 16–19 December, 1981, pp. 205–236 (Calcutta: Indian Statistical Institute).

Diaconis, Persi and Sandy Zabell (1982) "Updating subjective probability," *J. Am. Stat. Assn.* **77**, 822–830.

Diaconis, Persi and Sandy Zabell (1986) "Some alternatives to Bayes's rule," in Grofman and Owen, pp. 25–38.

151

Field, Hartry (1978) "A note on Jeffrey conditionalization," *Philosophy of Science* **45**, 361–367.

Freedman, David (1977) "A remark on the difference between sampling with the without replacement," *J. Am. Stat. Assn.* **72**, 681.

Freedman, David and Roger A. Purves (1969) "Bayes' method for bookies," *Annals of Math. Stat.* **40**, 1177–1186.

Garber, Daniel (1980) "Field and Jeffrey conditionalization," *Philosophy of Science* **47**, 142–145.

Goldstein, Michael (1983) "The prevision of a prevision," *J. Am. Stat. Assn.* **78**, 817–819.

Goldstein, Michael (1985) "Temporal coherence," in Bernardo et al. (1985), 231–248.

Good, I. J. (1981) "The weight of evidence provided by uncertain testimony or from an uncertain event," *Journal of Statistical Computation and Simulation* **13**, 56–60.

Grofman, Bernard and Guillermo Owen (eds.) (1986) *Information Processing and Group Decision Making* (Greenwich, Conn.: JAI Press).

Harper, William and C. A. Hooker (eds.) (1976) *Foundations of Probability Theory, Statistical Inference, and Statistical Theories of Science* (Dordrecht: D. Reidel Pub. Co.).

Jaynes, E. T. (1959) *Probability Theory in Science and Engineering* (mimeographed) (Dallas: Field Research Laboratory, Socony Mobil Oil Co.).

Jeffrey, Richard (1957) *Contributions to the Theory of Inductive Probability* (Ph. D. Dissertation, Princeton University).

Jeffrey, Richard (1965) *The Logic of Decision* (New York: McGraw-Hill Pub. Co.: 2nd ed., University of Chicago Press, 1983).

Jeffrey, Richard (1975) "Carnap's empiricism," in Maxwell and Anderson (1975).

Jeffrey, Richard (ed.) (1980) *Studies in Inductive Logic and Probability*, vol 2 (Berkeley and Los Angeles: University of California Press).

Jeffrey, Richard (1985) "Probability and the Art of Judgment," in Achinstein and Hannaway (1985), 95–126.

Kyburg, Henry E., Jr. and Howard E. Smokler (eds.) (1964) *Studies in Subjective Probability* (New York: Wiley. 2nd ed., Robert E. Krieger: Huntington, N. Y., 1980).

Lewis, C. I. (1946) *An Analysis of Knowledge and Valuation* (La Salle, Ill.: Open Court).

Maxwell, Grover and Robert M. Anderson (eds.) (1975) *Induction, Probability, and Confirmation* (Minneapolis: University of Minnesota Press).

Mellor, D. H. (ed.) (1980) *Prospects for Pragmatism* (Cambridge: The University Press).

Skyrms, Brian (1980a) *Causal Necessity* (New Haven: Yale University Press).

Skyrms, Brian (1980b) "Higher-order degrees of belief," in Mellor (1980), 107–137.

Skyrms, Brian (1984) *Pragmatics and Empiricism* (New Haven: Yale University Press).

Skyrms, Brian (1987) "The value of knowledge," in *Justification, Discovery and the Evolution of Scientific Theories*, C. Wade Savage (ed.) (Minneapolis: University of Minnesota Press).

Suppes, Patrick (ed.) (1980) *Studies in the Foundations of Quantum Mechanics* (East Lansing, Mich.: Philosophy of Science Association).

Suppes, Patrick and Mario Zanotti (1980) "A new proof of the impossibility of hidden variables using the principles of exchangeability and identity of conditional distributions," in Suppes (ed.) (1980), 173–191.

Teller, Paul (1976) "Conditionalization, observation, and change of preference," in Harper and Hooker (1976), pp. 205–259.

van Fraassen, Bas (1980) "Rational belief and probability kinematics," *Philosophy of Science* **47**, 165–187.

van Fraassen, Bas (1984) "Belief and the will," *J. Phil.* **81**, 235–256.

von Mises, Richard (1919) "Grundlagen der Wahrscheinlichkeitsrechnung," *Mathematische Zeitschrift* **5**, 52–99.

von Mises (1957) *Probability, Statistics, and Truth*, 2nd rev. ed. (London: George Allen and Unwin. Dover reprint, 1981).

Zaman, Arif (1984) "Urn models for Markov exchangeability," *Annals of Probability* **11**, 223–229.

8

Preference among preferences

In *The Logic of Decision*,[1] preference is represented as a relation between propositions. To the same effect, it could have been represented as a relation between sentences which express those propositions.[2] Represented in this way, the preference relation belongs to the same syntactical category as the relation ⊢ of logical implication. But matters become more complex when we write sentences (and not their names, and not names of propositions, either) beside the preference symbol "pref" to get sentences of form "*A* pref *B*" which may themselves flank that symbol.[3] The symbol is then a connective, not a two-place predicate. With its aid, one can express not only such first-order preferences as that George M. Akrates prefers smoking to abstaining ("*S* pref ~*S*"), but also such second-order preferences as that he prefers preferring abstaining:

$$(\sim S \text{ pref } S) \text{ pref } (S \text{ pref } \sim S).$$

First published by R. Jeffrey, in *The Journal of Philosophy*, Vol. 71, No. 13, pp. 377–391. Copyright 1974 by Columbia University.

I have been tinkering with higher-order preference intermittently since ca. 1969 – with support of the National Science Foundation, which is gratefully acknowledged. Earlier drafts of this paper have been presented at Chelsea College, The Johns Hopkins University, The University of North Carolina at Chapel Hill, and the University of Pennsylvania, generally under a title, "Preference and Modality," which I have abandoned against the better judgment of Fabrizio Mondadori. Apart from acknowledgments in the text, thanks are due to Jay Rosenberg, James Cornman, Edwin McCann, and, especially, Brian Chellas, for comments which have influenced the final version. I read Harry Frankfurt's paper (in 5) tardily, after finishing this. I take him to have been the first to have deployed what is essentially the notion of higher-order preference, in an ingenious account of free action, free will, and the concept of a person.

1. Richard C. Jeffrey (McGraw-Hill, 1965; University of Chicago Press, 1983, 1990).
2. If propositions are thought of as sets of logically equivalent sentences, then *expresses* coincides with the relation of membership between sentences and propositions. If propositions are thought of as sets of maximal consistent sets of sentences, then the proposition expressed by a sentence is the set of all maximal consistent sets of sentences to which that sentence belongs.
3. Observe that in a well-formed sentence "*A* ⊢ *B*," "*A*," and "*B*" must be *names* of sentences.

Nor need we stop here; e.g., we can make the third-order statement that, as between this second-order preference and his first-order preference for smoking, he prefers the former:

$$[(\sim S \text{ pref } S) \text{ pref } (S \text{ pref } \sim S)] \text{ pref } (S \text{ pref } \sim S).$$

Here, "pref" belongs to the same syntactical category as C. I. Lewis's symbol "⊰" for strict implication. In moving from preference as a relation to preference as a connective, we increase modal involvement to the same degree as in moving from the relation ⊢ of entailment to the connective ⊰ of strict implication.[4]

I take it that humans (barring infants, psychopaths, . . .) have higher-order preferences and that few if any other animals do, even though other animals may well have first-order preferences.[5] My late cat clearly preferred milk to water on various occasions, but that was surely the end of it. It is not that he was complacent about his preference for milk in a sense in which Akrates, above, is *not* complacent about his preference for smoking (when he prefers smoking, but prefers preferring abstaining). The point is rather that cats are not aware of their preferences in the way in which they are aware of saucers of milk – as objects of desire or aversion, which at least occasionally can be sought or avoided. They are not self-conscious or self-manipulative in that way.[6] But people are, or can be; and therefore in discussing human preferences we want the additional expressive power that is obtained if we manage to represent preference as a sentential connective, and not simply as a relation between sentences.[7]

Skepticism about the existence of higher-order preference – or, if you will, skepticism about the need to include higher-order preferences in accounts of human action and motivation – may reflect

4. Keeping propositions as the *relata* of preference, the increased modal involvement would be obtained by moving from preference as a relation to preference as a non-Boolean operation on propositions, taking pairs of propositions as arguments and propositions as values.

5. Cf. "Freedom of the Will and the Concept of a Person," this JOURNAL, LXVIII, 1 (Jan. 14, 1971): 5–20, where Harry Frankfurt takes the ability to form second-order desires to be the mark of personhood.

6. That is part of their charm. Dog-lovers have suggested to me that part of *their* (dogs') charm is a rudimentary self-consciousness about their preferences.

7. To treat animals as having preferences between (the truth of) sentences is not to treat them as language-users: see my *Logic of Decision*, §4.7.

   Note that the same increase in expressive power is obtained by treating preference as a proposition-valued operation on pairs of propositions, as suggested in fn 4.

155

awareness that for the most part we do not simply choose our preferences, any more than we choose our beliefs, so that direct manifestation of preference in choice is rare when the objects of preference are themselves preferences. But although Akrates cannot simply choose to prefer abstinence on this occasion – and that, after all, is why his preference for preferring abstinence is (uneasily) compatible with his preference for smoking – he can undertake a project of modifying his preferences over time, so that one day he may regularly prefer abstinence, just as now he regularly prefers smoking. The steps toward this desired end may involve hypnosis, reading medical textbooks, discussing matters with like-minded friends, or whatever. But in accounting for Akrates's undertaking of these activities it seems natural to cite his preference for preferring abstinence, just as in accounting for his activities as he flings drawers open and searches through pockets of suits, one may cite his preference for smoking – imagining, in this latter case, that here smoking is not something he can simply choose, because the means are not at hand: He has run out of cigarettes.

There is a traditional view of Akrates's conflict which sees it as a tug o'war between Appetite and Will. Appetite pulls toward smoking. If Akrates were a simple pleasure-seeker, Will would march arm in arm with Appetite, and there would be no conflict. But Akrates would be reasonable, and therefore Will pulls against Appetite. Being weaker than Appetite on this occasion, Will loses the struggle: Akrates (the taut rope) is drawn into smoking, against his will.

So decked out, as a little theory, the traditional way of speaking seems ludicrous. But there are situations in which it is apt to speak of weakness of will, and of strength of appetites for food, sex, cigarettes, or whatever, and where we aptly speak not of preferences but of needs, wants, wishes, and the like. Capitalized and sealed into a tight little language game, Appetite and Will appear to live only by taking in each other's washing; but as part of the full linguistic apparatus that we bring to bear in electing action and in understanding it, these notions have real uses. In particular, we speak of appetites, will, needs, wishes, etc. in making judgments about what people's preferences are.

In *its* tight little language game, the technical term "preference" lives by exchange of washing with the notions of choice, optionality, and judgmental probability. But to make practical use of these concepts, we must bring them into contact with ordinary talk of needs, wishes, preferences, etc., somewhat as we must be prepared to recognize the colors and odors and tastes of things in order to put theoretical chemistry to use. (To get started, we must be able to identify this particular odorless salty white stuff – at least tentatively – as NaCl.) The general idea is that wishes, needs, lusts, etc. are all prima facie signs of *pref* (= preference in the technical or regimented sense) and that, in case of conflict, *pref* is shown by the outcome, which need not be evident in action; e.g., because, of two sentences, that which is preferred true may not be in the agent's power to make true. For the most part, when a statement of form "*A* pref *B*" is thought to be true, the corresponding colloquial statement in terms of preference is taken to be clearly true (on a sufficiently fussy reading), even though it may be thought inappropriate: not what would spring to mind as apt under the circumstances. Perhaps the following adaptation of a remark of Dale Gottlieb's will be helpful here: For the most part, when someone acts in order to make *A* true, and in fact *A pref B,* and the only options are making *A* true and making *B* true, the statement "*A pref B*" is not satisfactory as a causal explanation-sketch of why the agent acted as he did. Rather, that statement is of some use in suggesting the form appropriate for such a sketch, but the sketch itself will mention needs, wishes, etc., which may be viewed as causes both of the *pref* and of the action. (Reasons, too, maybe.)

To a first approximation, free action is simply action in accordance with preference (cf. Frankfurt, op. cit.). In this sense, Akrates acts freely when he smokes because (= from motives in view of which) he then prefers smoking to abstaining, even though he then prefers preferring abstaining. To be a bit more precise, his smoking is then a free action as long as smoking and abstaining are both options for him, even if preferring to abstain is not. In contrast, consider the case in which Akrates smokes now out of compulsion – against his preference for abstaining (and against his will, in another way of speaking). Here the position is that abstaining is not an option for him, any more than smoking is an option for him when he can get no cigarettes. But perhaps (as Clark Glymour has suggested) action in accordance with preference is not free if the agent would undo that preference if he could. Here as throughout, my aim is not

157

to explicate such common terms as "freedom" and "compulsion" but to show how the notion of higher-order preference can draw no less real distinctions, to much the same effect, in its own terms. It seems to be only occasionally that preferences are optional, and these occasions may well be seen as fateful. It may well be that Akrates chose his preference for smoking at the age of 14. Perhaps he saw it as part of a larger option: for an adult, masculine, vivid style instead of something prudent or tame or prissy. But ten years and 100,000 cigarettes later, preferring abstention is no longer an option. Early on, each cigarette was a ratification of his choice of preference for smoking, but long since, these acts have become simple *expressions* of well-entrenched preference. Meanwhile, his second-order preferences may have changed, e.g., in response to data about the effect of smoking on health coupled with a sense of having been had, a decade earlier, by the cigarette companies. At this point in his life, Akrates's position would seem to be this:

$$S \text{ pref } \sim S, \quad (\sim S \text{ pref } S) \text{ pref } (S \text{ pref } \sim S)$$
$$OS, \quad O \sim S, \quad \sim O(\sim S \text{ pref } S)$$
$$S$$

where "O" may be read, "it is optional that." More needs to be said about optionality, soon.

Meanwhile, notice another way in which preferences may be chosen: the way of The Good Soldier who, in obeying an order, acts in accordance with a preference or a set of preferences which stem from the order and from his commitment to adjust his preferences to his orders, but from no wish or desire or need etc. of the sort that such action would be seen as expressing by someone who took the soldier to be acting on his own, and not in response to an order. The order may be to take a certain hill. There is no question of robotlike response, for the order is not a detailed set of instructions for the placement of his feet, etc. The Good Soldier will use his wits to make a series of decisions with a view to achieving the required objective, in the light of a background of preferences which may correspond to standing orders about circumstances in which it is acceptable to risk life and equipment in various categories. Adopting a preference on command may well be a matter of preferring in spite of contrary wishes, inclinations, appetites, and fears. Here one may speak of courage, or will power, depending on how smoothly the struggle is won – when it is won. Adopting a preference on command may also be a matter of setting aside standing preferences which, if questioned in civilian life, would be defended in terms of morality or common decency, and which, for that very reason, are not questioned there. Returning to his old life, the soldier may come to see his creditable military performance as acquiescence in monstrous evil.[8] I mention this in order to support the view that following orders is a matter of adopting preferences: The suggestion is that remorse is embittered by a sense of real complicity, for the good soldier acts freely, in accordance with first-order preferences he has freely adopted in accordance with a second-order preference for adopting certain sorts of first-order preferences on command.

If preferences are in the mind, the mind is not always transparent to the mind's eye: One may be unaware of what one's preferences are, in particular cases. One may be in doubt about what they are. And one may be in error in the matter. I take the touchstone to be choice: If A and B are options, if they are all the options, and if you believe both of those things, then choice reveals preference in that, choosing A, you cannot have preferred B. But this is far from being a behavioral or operational test. The soldier who, thinking he was about to obey the order to advance, finds himself in flight with empty bowels, may be judged (e.g., by a court martial) to have acted in accordance with a keen preference not to advance; or he may be judged to have acted under compulsion, against his preference to advance. (Will lost the tug o'war with Fear.) On the second reading, advancing was not an option, and the soldier's preference was as he had thought it. On the first reading, advancing was an option, but he was wrong about his preference. Unexpected sexual shrivelings are another case in point. Do not Casanova's elaborate preparations to bed the lady make it clear where his preference lies? If so, consummation was, unexpectedly, not an option for him. But on the other hand, does not his most basic indicator of present sexual preference give the lie to his feverish preparations, and show that his real preference is not what he had thought? In such cases,

8. "And you have to come home knowing you didn't have the guts to say it was wrong. A lot of guys had the guts. They got sectioned out, and on the discharge, it was put that they were unfit for military duty – unfit because they had the courage. Guys like me were fit because we condoned it, we rationalized it." *Time*, October 23, 1972, p. 34.

159

preference and optionality exhibit a sort of complementarity: In order to maintain belief in the one, it is necessary to deny the other.

Nor, in the absence of such ambiguities, need it be clear what preference underlies a publicly observable act. As Donald Davidson has emphasized,[9] actions are events, and events are concrete particulars. But preference relates propositions, not events. If the soldier preferred flight to attack, and fled, his flight is a concrete particular, having no end of properties. (One might describe the course of his flight, e.g., in unlimited detail.) But since he did not attack, there is no concrete particular ("his attack") to serve as the other term of the preference relation, here. When you act in accordance with your preferences, you enact a certain proposition: one of the highest-ranking among those of your options which you believe to be options. You enact the proposition A (or, equivalently, you make the corresponding sentence true) by performing some perfectly definite act – an act which can be described as making A true, but which can also be described truly as making any number of other propositions true. Among all these, it need not be obvious either to the agent or to an observer which is the proposition for the sake of the truth of which the act was performed. Here, some may prefer to speak of preferences and choices among *possible* acts, only one of which will prove to be *actual,* viz., the act chosen. But I prefer sentences or propositions as relata of preference. These have the virtue of clarity: Ordinary logic provides a satisfactory account of their interrelations, and on top of that account one can construct what I take to be a satisfactory account of preference. But *possible acts* remain to be clarified. I think the burden of clarification is on those who would construct an account of preference in such terms.[10]

If the system of *The Logic of Decision* is to be extended to

9. In numerous publications, e.g., in "The Logical Form of Action Sentences," in Nicholas Rescher, ed., *The Logic of Decision and Action* (Pittsburgh: University Press, 1967).

10. An example of such an account would seem to be that of L. J. Savage, *The Foundations of Statistics* (New York: Wiley, 1954): He takes the relata of preference to be certain entities which he calls "acts" and which might better be called "possible acts." These entities are functions from the set of all possible worlds to a set of entities called "consequences." The consequences may, but need not, be hedonic states of the agent. Here, one surely never knows just what act one is performing; for to know that, one would have to know how the act would turn out in every possible world. If possible acts are such as these, they cannot be the objects of choice. These points are elaborated in my "Frameworks for Preference" in M. Balch, D. McFadden, and S. Y. Wu, eds., *Essays on Economic Behavior under Uncertainty* (Amsterdam: North-Holland, 1974).

160

encompass preference among preferences, it is essential that matters stand as I have argued that they do: that one need not be aware of just what one's preferences are, and that one may have (and, commonly, does have) various degrees of belief between 0 and 1 in sentences of form "*A* pref *B*" where "pref" refers to one's own preferences. If, on the contrary, degrees of belief in such sentences always or commonly take only the extreme values, 0 or 1, depending on whether the sentences are false or true, there can be no interesting structure of higher-order preferences. The reason is that, in the system of *The Logic of Decision,* the agent must be indifferent between any two propositions in which he has full belief. It would follow that (always or commonly) the agent will be indifferent between pairs of propositions that truly describe his actual preferences. Thus, if he prefers smoking to abstaining, but prefers preferring abstaining, he will be indifferent between those two preferences, simply because he knows that they *are* his preferences.

We wish to leave open such possibilities as that

(*) $[(\sim\!S$ pref $S)$ pref $(S$ pref $\sim\!S)]$ pref $(S$ pref $\sim\!S)$

when both relata of the main occurrence of "pref" are true. We have seen that we can do this only if truth, in such cases, does not imply full belief: the statement (*) will be false – there will be indifference, not preference, between $(\sim\!S\ pref\ S)\ pref\ (S\ pref\ \sim\!S)$ and S *pref* $\sim\!S$ – if *prob,* the agent's judgmental probability function, assigns the value 1 to both sentences. But is this not too high a price to pay? Do we not wish to be able to assert (*) even when the agent is sure that his preferences are as described on the two sides of the main occurrence of "pref"? I think not. Here my reasons are the same as my reasons for holding that, in general, the agent must be counted as indifferent between the truths of *any* two sentences of whose truth he is utterly certain. Preference for truth of one sentence over truth of another can be thought of in a rough and ready way as willingness to pay something in order to have the preferred sentence come true, when the option is between truth of the one and truth of the other. But where the agent is quite sure to begin with that both sentences *are* true, he should not be willing to pay anything to have either of them *come* true. Thus, where the agent is quite sure that (S) he will smoke now, and equally sure that $(S$ pref $\sim\!S)$ he prefers smoking now to abstaining now, he must be indifferent between

161

these two. Preference is a practical concept which takes beliefs fully into account in this way. Of course, all of this is compatible with (say) present preference for future preference for abstaining then, over smoking then, where we use two different preference relations, and the sentence about smoking is no longer S. I shall have nothing to say in this paper about the influence of present preference and present action on future preference.[11]

With so much preamble, let us examine various hypotheses about Akrates's situation with a view to seeing how prima facie difficulties may be overcome.

We have noted that if abstention is not an option for Akrates, there is no difficulty in understanding how he can smoke and yet prefer not to. Here there is no need to invoke higher-order preferences in order fully to describe his situation.

But the case suggested at the beginning is one where Akrates smokes *and prefers to smoke,* but is dissatisfied with his smoking because (a) abstention is an option, and he knows it, and (b) he is dissatisfied with his first-order preferences, and wishes they were otherwise: He prefers preferring abstaining. It is tempting to insist that matters must stand as follows, where, again, OX means that X is an option for Akrates.

   (i) S pref $\sim S$
   (ii) S
   (iii) $(\sim S$ pref $S)$ pref $\sim(\sim S$ pref $S)$
   (iv) O $\sim S$
   (v) $\sim$O$(\sim S$ pref $S)$

The point of asserting (v) is that if preference for abstention were one of Akrates's options, his preference for that option over its denial would seem incompatible with his not taking it, i.e., with his actual preference for smoking over abstaining. But this reason for (v) will not do.

11. The present discussion of preference is purely synchronic. Thus, a third plausible analysis of the situations of Casanova and The Bad Soldier was omitted, above: They might have changed their minds!

In "Higher Order Probabilities and Coherence," *Philosophy of Science,* XL (1973): 373–381, Soshichi Uchii argues that incoherence results unless $P(P(h) = p)$ always has the value 1 or 0 depending on whether $P(h) = p$ or not, but (I think) begs the question by packing that conclusion into the very definition of *possible world:* See (iv) on his p. 376. If cogent, a corresponding argument with "A pref B" in place of "$P(h) = p$" above would show the present theory of higher-order preference to be trivial.

The difficulty is that there may be incompatible options, each of which Akrates prefers to its denial. Then the principle that each option which is preferred to its denial must be true would lead to the conclusion that Akrates makes a logical impossibility true.[12] The rejected principle is this:

$$\frac{OX}{X \text{ pref } {\sim}X}{X}.$$

Thus, suppose that A and B are sentences which Akrates would like to have agree in truth value – both true or both false, he cares not which – and that both $A \& B$ and ${\sim}A \& {\sim}B$ are options. Setting $X = A \& B$ in the rejected principle, we have A and B both true, and then setting $X = {\sim}A \& {\sim}B$ there we also have A and B both false. Concretely, let $A = I$ shall study Professor Moriarty's treatise on the binomial theorem this afternoon and let $B = $ This evening I shall try to explain the binomial theorem to Dr. Watson, where the laborious business of enacting A would so enhance the probability of success of B as to make $A \& B$ an attractive option (preferred to its denial), and where it would also be attractive to escape both tasks.

The solution to this difficulty – so I think – is to compare X with all other options, and not simply with ${\sim}X$ (when that happens to be an option, which it need not be). Thus, an improvement over the rejected principle would be this, in idiomatic English[13]:

A sentence must be true if making it true is an option preferred to every other.

It is simply not good English to speak of sentences as options, although I shall continue to do so for fluency when the sentences are represented in logical notation. "OX" is best read, "Making X true is an option," and that may be taken as the intent of my barbarous "X is an option."

---

12. One might try to avoid the difficulty by drastically weakening the principle, so as to require only that an option be false if its denial is preferred to it. The underlying thought ("the Anscombe-Kanger principle") is that good is good enough. I prefer another way out, in which connections among preference, optionality, and action remain relatively tight; see below.

13. The improved principle is obtained from the rejected one (a rule of inference) by replacing the second premise by this: $(Y)\{[{\sim} \Diamond (X \& Y) \& OY] \rightarrow X \text{ pref } Y\}$.

To a first approximation, an option is a sentence which the agent can be sure is true, if he wishes it to be true. (Thus, $A \lor \sim A$ is always an option, trivially.) But there may be options in this sense which the agent does not know are options, as when he fails to realize that a door is unlocked, or that his car will start, or that his love is reciprocated, or that if he holds out one more day, the pain of withdrawal will begin to ease. Perhaps such unrecognized options should not be regarded as options at all, or perhaps some should but others should not. At any rate, the principle seems to need further restatement, perhaps as follows:

A sentence is true if making it true is a recognized option preferred to every other recognized option.

I shall continue to use the term "option" in the broad sense in which not every option need be recognized as such. Problem: How shall we characterize the recognized options? Can this be done in the terms we have already deployed, or is a new primitive term needed? One might think of defining recognition as full belief in a truth: Where *prob* is the agent's judgmental probability function, one might take joint truth of OX and prob(OX) = 1 to be necessary and sufficient for X to be a recognized option. The question whether prob(OX) = 1 can be settled by examining the agent's preference ranking, in which such sentences as OX and $O \sim X$ are presumed to occur. (For details, see my *Logic of Decision*, §7.4.) On this reading, recognition is definable in terms already at hand, but it is unclear that this is an adequate reading. Must one be as sure that OX as one is that $2 + 2 = 4$ in order to be said to recognize that making X true is an option? The basic difficulty here is a familiar one: lack of fit between ordinary talk of knowledge, belief, and recognition on the one hand, and such notions as judgmental probability and preference on the other. The principle we have been elaborating becomes a rule of very narrow scope when recognition is defined in terms of *prob* as above. For the record, it looks like this:

$$\frac{OX \ \& \ \text{prob}(OX) = 1 \quad (Y)([\sim \Diamond (X \ \& \ Y) \ \& \ OY \ \& \ \text{prob} \ (OY) = 1] \rightarrow X \text{ pref } Y)}{X}.$$

164

To have such a rule of inference, the theory of preference and optionality must have abundant logical resources, even at the level of propositional logic.[14] Thus, it must have the means to quantify over propositions; it must contain a notation for possibility; and it must have a notation for full judgmental probability. All of this must be available in the object language. On the other hand, if we use the colloquial form of the rule, giving "recognized" its colloquial reading, we have a looser principle of broad scope, extraneous to the theory of preference, but serving to link that theory with ordinary ways of speaking. Perhaps the principle is best taken in its colloquial form. One might still include optionality as a concept within the theory of preference, and might use the colloquial principle as a guide in dealing with particular applications of the theory.

We have roughly characterized an option as a sentence which the agent can be sure is true, if he wishes it true. Presumably the reference to wishing, here, can be replaced by a reference to preference. Can we then define optionality in terms of preference? Any such definition would seem to require consideration of whether the optional sentence would be true if the agent's preferences and beliefs were other than they are. The Small Girl in Lewis Carroll's joke has taken a faltering step in that direction[15]:

"I'm so glad I don't like asparagus," said the Small Girl to a Sympathetic Friend, "Because, if I did, I should have to eat it – and I can't bear it!"

Such a definition has been suggested by David Lewis, using the notion of a *partition*, viz., a set of sentences, no two of which can be true and at least one of which must be true. Such a set is a *partition of options* if and only if the following holds for each member: If it were preferred to every other member then it would be true. Now a *basic option* is defined as any member of any partition of options, and an *option* is defined as anything implied by a basic option.

This definition of "O" will serve if each option really is implied

---

14. That is the level, throughout this paper. Extension to first-order logic seems possible, using methods developed by Jerzy Łoś ["Remarks on the Foundations of Probability," *Proceedings, International Congress of Mathematicians*, Stockholm (1962): 225–229] and Haim Gaifman ["Concerning Measures on First-order Calculi," *Israel Journal of Mathematics*, II (1964): 1–18].
15. Quoted by W. W. Bartley III in *Scientific American*, 227 (1972): 39.

165

by some basic option; and that seems to be the case. Thus, one of my options is dying before the age of fifty years, e.g., by my own hand, but that is no basic option, for, I regret to say, its denial ("Live at least fifty years") is nothing I can ensure. But it is an option according to Lewis's definition: There is a partition of options ("Jump out the window"; "Don't"), one member of which implies it. Lewis's analysis seems right. I am still a bit queasy about counterfactuals, but Lewis is not, and perhaps he is right, not to be.[16]

The problem of providing a semantics for higher-order preference seems fairly straightforward, when optionality is bracketed. The basic point (suggested by David Lewis ca. 1966) is that nothing in the system of *The Logic of Decision* requires us to suppose that the terms of the preference relation are naturalistic propositions. They may equally well be propositions about the agent's preferences. In the case of propositional logic with *pref* and *ind* (for "indifference") as the only non-truth-functional connectives, the set L of sentences under consideration may be taken to be the closure of a countable set of atomic sentences under the connectives $\sim$, &, *pref*, and *ind*. A *model* of L is determined by the following five items:

(1) A nonempty set W of "possible worlds"
(2) A Boolean algebra of subsets of W, viz., "propositions"
(3) For each w in W, a probability measure P_w on the algebra
(4) For each w in W, a ("utility") function u_w which assigns a real number $u_w(w')$ as value to each argument w' in W in such a way that the conditional expectation $E_w(u_w \mid A)$ of u_w on A relative to

---

16. See his *Counterfactuals* (Oxford: Blackwell, and Cambridge, Mass.: Harvard, 1973). The following curious feature of Lewis's definition of optionality was pointed out by Joseph Ullian when the present version was read at MIT. Consider the second of the three preference rankings displayed in the penultimate paragraph, below, and abbreviate "S pref $\sim S$" by "P." Since $\{P, \sim P\}$ is a partition and both P and $\sim P$ are options, one might expect $\{P, \sim P\}$ to be a *partition-of-options* in Lewis's technical sense; but it cannot be, for $\sim P$ is false although it is preferred to P. To this Lewis replies that P and $\sim P$ cannot be *basic* options in this case. Instead, the basic options must be $S \& P$, $S \&$ $\sim P$, $\sim S \& \sim P$, and $\sim S \& P$, in descending order of preference. But, as Lewis conjectures, the situation remains disturbing in that although the basic option $S \& P$ is actualized, Akrates cannot clearly realize that, for one can show that the given preference ranking is possible only when Akrates's degree of belief in $S \& P$ is less than $1/2$. [For P to be at the bottom of the ranking, the probability-weighted average of the best and worst basic options must be less than that of the two lowest. From this, with some labor, one can get prob($S \& P$) $<$ prob($\sim S \& \sim P$), whence the result follows.]

the probability measure P_w exists for each A to which that measure assigns positive probability[17]

(5) For each S in L, an element $M(S)$ of the Boolean algebra ("the proposition expressed by S") satisfying the conditions:

(a) $M(\sim S) = W - M(S)$

(b) $M(S \& T) = M(S) \cap M(T)$

(c) $M(S \text{ pref } T)$ = the set of all those worlds w which satisfy the inequality $E_w(u_w \mid M(S)) > E_w(u_w \mid M(T))$, where the inequality is taken to fail if either side is undefined

(d) $M(S \text{ ind } T)$ = the set of all those worlds w which satisfy the condition $E_w(u_w \mid M(S)) = E_w(u_w \mid M(T))$, which is taken to fail if either side is undefined

I am indebted to Zoltan Domotor for help in getting this right, if it is right.

The heart of the construction is in (5)(c) and (d), which ensure that "S pref T" (or "S ind T") will be true in world w if and only if S is higher than T (or S is at the same level as T) in the preference ranking determined by the agent's values (u_w) and beliefs (P_w) as they would be in that world. In general, S is true in world w if and only if w belongs to $M(S)$. Truth *tout court* is truth in the real world, viz., an unknown element r of W.[18] Validity in a model is truth in every member of the set W for that model. Universal validity is validity in every model. In terms of this semantics one might discuss probabilities of probabilities, probabilities of preferences, etc., but this is not the place for it.

To conclude on a less technical note, let us examine a set of preferences more contorted than any attributed to Akrates heretofore. Suppose that Akrates's preference for smoking, like the smoking itself, is the outcome of a preferential choice, and that nevertheless,

17. In computing the conditional expectation, w is held fixed: $E_w(u_w \mid A)$ is the probability-weighted average of the values $u_w(w')$ that the function u_w assigns to arguments w' in A, where the weights are determined by the fixed probability measure P_w. The resulting conditional expectation is a function of w but not of w'. Thus,
$$E_w(u_w \mid A) = (1/P_w(A)) \int_A u_w dP_w.$$
18. One might think of adding a sixth item to the five that are taken to determine a model: (6) A distinguished member r of W which satisfies the condition that for each S in L to which P_r assigns positive probability, $E_r(u_r \mid M(S))$ is the P_r-weighted average of the values assumed by $E_w(u_w \mid M(S))$ as w ranges over W. But perhaps this condition, if apt, should be imposed on every element r of W.

167

he prefers not preferring to smoke. Thus, each of the following is true:

$$S, \quad S \text{ pref } \sim S, \quad \sim(S \text{ pref } \sim S) \text{ pref } (S \text{ pref } \sim S).$$

Furthermore, the first two of these propositions are optional, as are their denials. Finally, Akrates believes (even, with judgmental probability 1) that these four are options and are all the options. Can we consistently suppose all this? I think so.

An air of paradox lingers around the claim that S pref $\sim S$ is true and optional, whereas its denial is preferred to it; but those are not all the options. Akrates's situation is intelligible enough if we inquire into the relative positions of S and $\sim(S$ pref $\sim S)$ in his preference ranking, and find that the former is higher, e.g., because we have one of the following two configurations, in which better-liked options are higher:

$$
\begin{array}{ll}
S & S \\
\sim(S \text{ pref } \sim S) & \sim(S \text{ pref } \sim S) \\
S \text{ pref } \sim S & \sim S \\
\sim S & S \text{ pref } \sim S.
\end{array}
$$

In each case, the fact that S is above $\sim S$ shows that "S pref $\sim S$" is true, even though the denial of that option is preferred to it. In each configuration, the top entry is true perforce, being the best of the (recognized) options. But in each case, one of the other entries is also true, and in neither case is that one the best of the remaining options. (In the second case, it is the worst that is chosen, along with the best!) It might seem plausible to explain the consistency of the first configuration by pointing out that there, the intensity of Akrates's preference for smoking over abstaining is greater than that of his preference for not preferring smoking over preferring it. But that need not be the case in the second configuration. The common, telling feature is simply that his desire to smoke is at least as intense as his desire not to prefer smoking to abstaining: In fact, S pref $\sim(S$ pref $\sim S)$. Were *that* preference reversed, inconsistency would ensue, e.g., in the following ranking:

$$
\begin{array}{l}
\sim(S \text{ pref } \sim S) \\
S \\
\sim S \\
S \text{ pref } \sim S.
\end{array}
$$

168

If these four are taken to be options and to be all the options, the topmost must be chosen; but then the order of the middle two cannot be as shown.

Perhaps enough has now been said, to indicate the nature and objectives of the theory of higher-order preference. Like the theory of first-order preference, it simply refuses to countenance as preferences rankings that are intransitive or have certain other failings – human though those failings be.[19] But the important point is that the higher-order theory does countenance various other failings – or misfortunes, or conflicts, or tensions, or "contradictions" in some Hegelian sense. It gives us a canvas on which to paint some very complex attitudinal scenes, from life. That one cannot paint intransitive preference rankings on that canvas makes it all the more interesting that one *can* paint poor Akrates there, in the various postures we have seen above.

19. My thought is that people who say they prefer A to B and B to C but not A to C are simply mistaken about their preferences. Others think that the theory imposes an idealization, and is false of much actual preference. Still others (e.g., Amélie Rorty) hold that intransitivities may be quite in order, e.g., when A is preferred to B in one respect (or, under one description) and B is preferred to C in (or, under) another. But throughout, I am concerned with preference *all things considered*, so that one can prefer buying a Datsun to buying a Porsche even though one prefers the Porsche qua fast (e.g., since one prefers the Datsun qua cheap, and takes that desideratum to outweigh speed under the circumstances). *Pref* = preference *tout court* = preference on the balance.

9

On interpersonal utility theory

Utility is a technical concept and a theoretical one, undiscussible in isolation from the theories in which it has its being. No more can the possibility of interpersonal comparison of utilities be discussed in isolation from theories in which such comparisons play a role. Thus, the problem is undiscussible where the theoretical context consists solely of the von Neumann–Morgenstern[1] theory of personal preference, for nothing there depends on the possibility of interpersonal comparison of utilities. *Nor does anything in the von Neumann–Morgenstern theory exclude the possibility of such comparisons.* The contrary impression, which is widespread, is based on the unfounded assumption that the theory of personal decision making is *the* definitive context for the utility concept.

In this paper I consider the problem of interpersonal comparison of utilities in the context of a theory (the New Utilitarianism) which is obtained by substituting the von Neumann–Morgenstern utility concept into (say) Bentham's normative political theory. In that context, interpersonal comparisons of utilities become interpersonal comparisons of preferences: "Individual i's preference for A over B exceeds (or is exceeded by, or is the same as) individual j's preference for C over D."

It is my contention that, commonly enough, we do make such intercomparisons, and that the theoretical apparatus of the New Utilitarianism can serve to test and to supplement them by bringing

First published by R. Jeffrey, in *The Journal of Philosophy*, Vol. 68, No. 20, pp. 647–656. Copyright 1971 by Columbia University.

This paper is a handily separable, representative part of a larger study which has been supported financially by The National Science Foundation, The John Simon Guggenheim Memorial Foundation, and The University of Pennsylvania. Thanks are due to the Philosophy Department of University College London and to the Philosophical Institute of Uppsala University for kind hospitality, room for work, and occasions for discussion.

1. John von Neumann and Oskar Morgenstern, *Theory of Games and Economic Behavior* (Princeton, N.J.: University Press, 1944; 2nd ed., 1947), §3.

them into confrontation with each other and with judgments about even-handedness of compromises. But this is no brief for the New Utilitarianism as a political or economic norm. It is enough for my purposes that, acceptable or not, the doctrine be intelligible.

I. THE NEW UTILITY CONCEPT

It must be observed that this theory is not to be identified with the psychology of the Utilitarians, in which pleasure had a dominating position. The theory I propose to adopt is that we seek things which we want, which may be our own or other people's pleasure, or anything else whatever, and our actions are such as we think most likely to realize these goods.[2] *F. P. Ramsey*

Having been in bad odor for decades, the utility concept was reconstrued and partially rehabilitated by von Neumann and Morgenstern (op. cit.), who first demonstrated to the learned world at large[3] that any sufficiently coherent preference ranking of a sufficiently rich set of prospects is describable by a function that attributes numbers to prospects in such a way as to satisfy the *expected-utility hypothesis*, viz.,

(1) The number assigned to a gamble is a weighted sum $px + qy$ of the numbers x and y that are assigned to the two incompatible possible outcomes, where the weights p and q are the probabilities of those outcomes.

In particular, in case $p = q = \frac{1}{2}$ (a "50–50 gamble," as on the toss of a coin), the number assigned to the gamble will be the straightforward average $\frac{1}{2}(x + y)$ of the numbers assigned to the possible outcomes.

We then have the following purely preferential way of comparing differences between the numbers, x, y, and v, w, which are assigned to pairs, X, Y, and V, W, of prospects:

(2) $x - y$ will be greater than or equal to or less than $v - w$ accordingly as a 50–50 gamble between X and W is ranked higher than or with or lower than a 50–50 gamble between Y and V.

---

2. Frank Plumpton Ramsey, *The Foundations of Mathematics and Other Logical Essays* (New York: Harcourt Brace, 1931), p. 173.

3. The same point had been demonstrated by Ramsey (op. cit.) in a (then) little-noticed paper, "Truth and Probability."

Proof. The condition $\frac{1}{2}(x + w) \geq \frac{1}{2}(y + v)$ is equivalent to the condition $x - y \geq v - w$. (Multiply both sides by 2 and transpose.)

A function that satisfies the expected-utility hypothesis (1) is called a *utility function*. A function that assigns the higher number to the higher ranked of two prospects and assigns the same number to each of two prospects that are ranked together is said to *represent* the ranking. Many different utility functions may represent the same preference ranking. But the proof of (2) assumed only that x, y, v, and w were the numbers assigned to the prospects X, Y, V, and W by *some* utility function which represented the ranking there. Consequently, (2) is true independently of the particular utility function that was used. This entitles us to define *inTRApersonal comparisons of preferences* as follows:

(3) An individual's preference for X over Y is greater than or equal to or less than his preference for V over W according as he prefers a 50–50 gamble between X and W to a 50–50 gamble between Y and V or is indifferent between the two gambles or prefers the second to the first.

Not every preference ranking is representable by a function that assigns numbers to prospects.[4] But von Neumann and Morgenstern (1947, appendix) proved that every preference ranking that meets certain conditions is representable *by a utility function*.[5] It follows that any such ranking is represented by an infinity of functions, e.g., by the function that assigns the value x^3 whenever the given utility function assigns the value x, and, in general, by any function that assigns the value y whenever the given utility function assigns the value x, provided the graph of y against x slopes up to the right.

4. EXAMPLE: The prospects are the points in a vertical plane. All the points in any horizontal line are preferred to any point below that line; and of two points on the same horizontal line, the one on the right is preferred. No assignment of real numbers to points can represent this ranking, for preferences between points on the same horizontal line are "infinitesimal" (but not zero) when compared with preferences between points on different horizontal lines.

5. For a simple exposition of these conditions, and the proof, see Herman Chernoff and Lincoln E. Moses, *Elementary Decision Theory* (New York: Wiley, 1959), pp. 79–85 and 350–352.

172

But for the most part, these new functions will not be utility functions; i.e., they will violate the expected-utility hypothesis.

Yet, infinitely many of the functions that represent the preferences of an individual i *will* be utility functions. The graph of the values assumed by any one of these against the values assumed by another will be a straight line, of the form $y = a_i x + b_i$ with a_i positive (ensuring that the line slopes up to the right):

(4) $\qquad u_i(A) = a_i w_i(A) + b_i \quad$ (a_i positive).

And, looking at it the other way around, if w_i is a utility function that represents a certain preference ranking, then so is u_i if (and only if) u_i is related to w_i by an equation of form (4).

II. THE NEW UTILITARIANISM

The community is a fictitious *body,* composed of the individual persons who are considered as constituting as it were its *members.* The interest of the community then is – what? The sum of the interests of the several members who compose it.[6] *Jeremy Bentham*

Bentham's doctrine was radically individualistic and egalitarian: He viewed society as a pseudo-individual, Leviathan, whose level of well-being is simply the aggregate of the levels of well-being of the constituent real individuals, where the mode of aggregation is simple summation. The obvious way of connecting these ideas with the new utility concept is to suppose that for each individual i there is a *welfare function* w_i which is in fact identical with one of the utility functions that represent his preferences. The number $w_i(A)$ represents the level of well-being that i expects to have in the event that prospect A is realized.[7] We then define Leviathan's welfare function, w_0, by the equation

(5) $\qquad w_0(A) = \Sigma_i w_i(A)$

which is asserted for every social prospect A.[8]

---

6. *An Introduction to the Principles of Morals and Legislation* (Oxford, 1789).
7. $w_i(A)$ will be a weighted average $p_1 x_1 + p_2 x_2 + \ldots$, where the x's represent levels of i's well-being in the various significantly different ways in which i thinks A might be realized, and the p's are the probabilities i attributes to these different modes of realization, conditionally upon A's being realized at all.
8. The index "i" ranges over the (real) individuals, who may have different sets of prospects; but each of Leviathan's prospects is also a prospect for each individual; i.e., it

173

The neo-Benthamite aggregation scheme (5) has natural anti-egalitarian generalizations of form

(6) $\qquad w_0(A) = \Sigma_i c_i w_i(A)$ $\quad (c_i$ positive)

where c_i is a constant that represents the degree to which i's well-being is taken into account in computing social welfare.[9] One might think of c_i as a measure of i's political power *de jure*.[10]

More sweeping generalizations are obtained if we allow that not all individuals need have the same legitimate interest in every social prospect. Here, c_i will be a function capable of assigning different numbers $c_i(A)$ to different prospects A, and the aggregation schemes are of form

(7) $\qquad w_0(A) = \Sigma_i c_i(A) w_i(A)$ $\quad (c_i(A)$ nonnegative).

These schemes cannot adequately be discussed in the space available here. I mention them now because, unlike schemes of form (6), they use the welfare concept at nearly full strength. Thus, when we replace the w_i of (7) by other utility functions u_i which represent the same individual preference rankings, the resultant social-welfare function cannot generally be expected to represent the same social preferences as those represented by the w_0 of (7) *unless the u_i are obtainable simply by multiplying the w_i by a common scale factor.*[11]

But schemes of form (6) will yield the same social preferences if we replace the welfare functions w_i by other utility functions u_i as in (4), *provided only that the scale factors a_i all be the same.* Adding different constants b_i to the different individual w_i merely has the effect of adding a constant $b_0 = \Sigma_i b_i$ to w_0, leaving the social preference ranking undisturbed. Then, for purposes of social decision making, there is nothing to choose between an aggregation scheme of form (6) and any aggregation scheme that defines social utility u_0 in the form

(8) $\qquad\qquad u_0(A) = \Sigma_i c_i u_i(A)$

must appear somewhere or other in each individual's preference ranking. (Not all of i's prospects need be acts; and some of i's prospects may be acts of others or of groups.)

9. If c_i were 0 for some i, that individual would not be a constitutive member of Leviathan. Negative values of c_i are excluded on the ground that it should not be part of Leviathan's aim to frustrate any of its members.

10. EXAMPLE: Leviathan is a commercial corporation; the individuals are the stockholders; and c_i is the number of shares that i owns.

11. Suppose in addition that $\Sigma_i c_i(A)$ is independent of A. Then the transformations of form (4) that preserve social preferences in (7) are those in which the a_i are all the same and the b_i are all the same; otherwise, the b_i must all be 0.

174

where the u_i are any utility functions related to the w_i by conditions of form

(9) $u_i(A) = aw_i(A) + b_i$ (a positive).

An assignment of utility functions u_i to the various individuals i will be called *interval-commensurate* when conditions of form (9) hold.

Now we can formulate the generalized neo-Benthamite aggregation scheme without using the notion of welfare: (6) may be replaced by (8) together with the condition that the u_i be interval-commensurate, and we can define interval-commensuration in terms of interpersonal comparisons of preference:

(10) Utility functions u_i and u_j are interval-commensurate if and only if i's preference for X over Y is greater than or less than or equal to j's preference for V over W accordingly as $u_i(X) - u_i(Y)$ is greater than or less than or equal to $u_j(V) - u_j(W)$.

Note that by (2), this accords with our definition (3) of inTRApersonal comparisons of preference; for (10) classifies u_i as interval-commensurate with itself. We have yet to examine the content of the notion of inTERpersonal comparisons of preference, although we shall, directly. Meanwhile, we can satisfy ourselves that (10) is equivalent to (9) as a characterization of interval-commensuration if we suppose that interpersonal comparisons of preference are related to welfare as follows:

(11) Individual i's preference for X over Y is greater than or less than or equal to j's preference for V over W accordingly as $w_i(X) - w_i(Y)$ is greater than or less than or equal to $w_j(V) - w_j(W)$.

Proof: By (9), the conditions $w_i(X) - w_i(Y) \geq w_j(V) - w_j(W)$ and $u_i(X) - u_i(Y) \geq u_j(V) - u_j(W)$ are equivalent (since a is positive).

Conclusion: In formulating the generalized neo-Benthamite aggregation scheme we can drop references to cardinal welfare as in (6) in favor of references to interpersonal comparisons of preference. Our final formulation of the New Utilitarian political norm is this:

175

(12) Social preferences ought to be those represented by a function u_0 that is related by some equation of form (8) to utility functions u_i which are interval-commensurate in the sense of (10).

Of special interest will be the equation of form (8) in which $c_i = 1$ for each i, viz., the egalitarian scheme

$$(13) \qquad u_0 = \Sigma_i u_i(A).$$

III. INTERPERSONAL COMPARISON
OF PREFERENCES

The New Utilitarianism makes sense as a norm – although the norm may be morally unacceptable or unworkable in practice – if and only if we can make sense of interpersonal comparisons of preferences, i.e., judgments of the form, "Individual i's preference for X over Y is the same as (or greater than, or less than) individual j's preference for V over W." Indeed we seem to make such judgments commonly enough, and back them up with reasons, as when we point out that i is willing to work harder to get X instead of Y than j is to get V instead of W. Such reasons are rarely conclusive, e.g., hard work is only one among many prima facie indicators of strength of preference, and it is entirely possible that these indicators may conflict. And the conflict may be of such a character as to defeat final judgment, although it may be resolvable by appeal to explicit or tacit criteria for weighing certain prima facie indicators more heavily than others in the presence of certain secondary data.

The New Utilitarianism provides a theoretical context within which such judgments can be tested against each other and against other judgments of a familiar sort, about the even-handedness of compromises. An important part of this theoretical context is the result of cumulative work over the past 20 years by Marcus Fleming,[12] John C. Harsanyi,[13] and Zoltan Domotor.[14] Thus, Harsanyi's work allows us to derive a complete set of interpersonal comparisons of preferences from a social preference ranking which is

12. Fleming, Marcus, "A cardinal concept of welfare," *Quarterly Journal of Economics* 66 (1952) 366–384.
13. Harsanyi, John C., "Cardinal welfare, individualistic ethics, and interpersonal comparisons of utility," *Journal of Political Economy* 63 (1955) 309–321.
14. Domotor's work on this topic is unpublished.

judged to be fair, together with the personal preference rankings of the constitutive individuals. The basic result[15] is a simple, complete characterization of the conditions under which a given social preference ranking is related to given individual preference rankings compatibly with New Utilitarian aggregation schemes of form (8), e.g., the egalitarian scheme (13)[16]:

(14) **Functionality.** *If everyone is indifferent between two prospects, so is society.*

(15) **Positivity.** *Society prefers one prospect to another whenever someone does and everyone else is indifferent between them.*

Suppose, now, that we are given social and individual preference rankings, all of which satisfy the von Neumann–Morgenstern conditions (note 5); and suppose, further, that we are given definite values for the coefficients c_i in (8). Then (note 15)

(16) Conditions (14) and (15) are jointly necessary and sufficient for the existence of utility functions u_0 and u_i which represent the social and individual rankings and satisfy (8) with the given c_i.

Note that (16) holds irrespective of the particular values assigned the c_i: suitable utility functions will exist in any case, but *which* particular ones will suit a particular case depends on the given c_i as well as the given rankings.

Normally, the functions u_0 and u_i of (16) are determined by the given constants c_i and the given preference rankings. Thus, Harsanyi (III) considers cases in which there exists a prospect Z and, for each individual i, a prospect U_i, where

(17) Individual i prefers U_i to Z, but everyone else is indifferent between U_i and Z.

15. Harsanyi (op. cit.) suggests a closely related result (see his fn 11) and proves a general form of the present result under assumption (17) below. I learned (16) from Domotor. See Peter C. Fishburn, "On Harsanyi's utilitarian cardinal welfare theorem," *Theory and Decision* 17 (1984) 21–28.
16. These conditions are noted by Fleming, op. cit., Postulate D. Note that scheme (7) violates (14) and (15).

Where there are just two individuals, this is equivalent to the condition that the individual preference rankings are neither exactly the same nor exactly the opposite.[17] Where (17) holds, we have the

(18) Uniqueness theorem (Harsanyi). *Any two sets u_0, u_i, and u_0, u_i, of utility functions which represent the given rankings and satisfy (8) with given coefficients c_i must be related by transformations of form $u_0'(A) = au_0(A) + b_0$, $u_i'(A) = au_i(A) + b_i$, where a is positive and the constants a, b_0, and the b_i are uniquely determined by the two sets of utility functions.*

Thus, under condition (17), the functions in the set u_0, u_i, will be interval-commensurate if and only if those in the set u_0', u_i' are.
Proof: See (9). It follows that under condition (17),

(19) A complete set of interpersonal comparisons of preferences is uniquely determined by (a) a set of preference rankings which satisfy (14) and (15), together with (b) a set of values for the coefficients c_i in (8).

Proof: See (10).

Except for certain special sorts of cases (e.g., see fn. 10) it is the egalitarian form (13) of aggregation scheme (8) on which we have the readiest intuitive grip, by way of our judgments of even-handedness of compromises. Here as in the case of interpersonal comparisons of preferences, prima facie indicators may conflict so as to defeat judgment; but often enough, conflicts are either absent or resolvable by appeal to secondary data, and judgments about even-handedness [viz., judgments that scheme (13) is operative] can yield interpersonal comparisons of preferences via (19).

For simplicity, consider the case where "society" consists of just two individuals, whose preferences are neither exactly the same nor exactly opposite. Suppose we have a social preference ranking that satisfies (14) and (15) and is judged even-handed in its treatment of the two individuals. Suppose, further, that the first individual is

---

17. Here, as throughout, we assume that the rankings satisfy the von Neumann–Morgenstern conditions.

178

indifferent between prospects V and W, and that the second is indifferent between prospects X and Y. Then

(20) The first individual's preference for X over Y is to be judged greater than or less than or equal to the second individual's preference for V over W according as society prefers a 50–50 gamble between X and W to a 50–50 gamble between Y and V, or has the opposite preference, or is indifferent between the two gambles.

Proof. By (3), relative to "individual" 0, together with (13).

IV. CONCLUDING REMARKS

There is much more to be said, and no hope of saying it in the space available here. But perhaps some brief remarks will aid discussion and understanding.

1. Boundedness and absolute commensuration. It is sometimes proposed, e.g., by Isbell,[18] that, since the St. Petersburg paradox and other considerations suggest that individual utility functions must be bounded above and below,[19] we effect a natural sort of commensuration, using no data beyond those available in the individual preference rankings, by adopting common upper and lower limits (say, 1 and 0) for the ranges of the functions u_i. But on the present account of these matters it is an empirical question, whether the commensuration obtained in this way agrees with other, commonly made interpersonal comparisons of preference and judgments of even-handedness which have their own claims to naturalness.

2. Identical subrankings and absolute commensuration. Whether or not individual utilities are bounded, one might form hypotheses about commensuration on the basis of features of individual preference rankings as follows. Suppose there is a large subset of prospects that are not gambles (e.g., a continuum of possible incomes)

18. J. R. Isbell, "Absolute Games," in A. W. Tucker and R. D. Luce, eds., *Contributions to the Theory of Games*, IV (Princeton, N.J.: University Press, 1959).
19. But see my *The Logic of Decision* (New York: McGraw-Hill, 1965), §10.2.

on which two individuals' preferences coincide. HYPOTHESIS: The commensurate utility functions for the two individuals will be those which assign the same values to all prospects in the subset. Where such hypotheses are plausible for two or more disjoint subsets, conflict becomes possible and therefore agreement becomes significant.

3. Theoretical contexts for absolute commensuration are provided by schemes of form (7). Thus, suppose there are just two individuals, viz., i and j; that $c_i(A) = c_j(B) = 1$ (say); and that $c_j(A) = c_i(B) = 0$. Then "society" prefers A to B if and only if i's expectation of welfare in case A exceeds j's expectation of welfare in case B.

4. The interpersonal analogue of preference. It can be shown (and will be, elsewhere) that absolutely commensurate utility functions (= welfare functions) can be derived from *interpersonal preference rankings* which meet certain conditions suggested by the correspondence between the inequality $w_i(A) \leq w_j(B)$ and the statement that i likes A no more than j likes B, where the "likes" refer to levels of well-being that i and j expect to ensue from A and B, respectively.

5. The factitiousness of preference. I think that interpersonal comparisons of preference are not significantly more difficult or methodologically tenuous than the inTRApersonal kind, but that the various idealizations about preference rankings on the basis of which one obtains individual utility functions may be more aptly described as falsifications when commensuration is in view. These idealizations constitute a sort of Logic of Decision which individuals can use as an anvil against which to form and reform parts of their preference rankings. But no actual preference ranking of more than a derisory fragment of one's prospects ever meets those standards. This fact is tolerable enough when the standards are used normatively (as an anvil) but not when (as in the New Utilitarianism?) they purport to describe existing preferences.

6. Fact versus value. Interpersonal comparisons of preferences determine and are determined by judgments about even-handedness of treatment of individuals in groups. But the intercomparisons are not therefore valuational, e.g., because even-handedness is not always desirable: Schemes (6) and (7) are sometimes apt.

7. On ceremony. Interpersonal comparisons of preference are connected with other factual judgments outside the confines of the theory I have been calling "The New Utilitarianism." Thus, i's willingness to lift and carry more to secure X instead of Y than j is to secure V instead of W is prima facie evidence that i's preference exceeds j's, here; but judgment should consult a variety of other facts as well, e.g., physiological facts about i's and j's musculature, sociological facts which might lead us to expect one of them to shirk the kind of labor that is required, etc. It would be futile to try to marshall these considerations into an explicit definition or into a comprehensive set of reduction sentences, but fatuous to think such exercises necessary steps in a process of elevating interpersonal comparisons of preference to the status of factual judgments.

10

Remarks on interpersonal utility theory

I. AN EXAMPLE OF INTERPERSONAL COMPARISON OF PREFERENCE INTENSITIES

Problem. Shall we open the can of New England clam chowder or the can of tomato soup, for the children's lunch? Adam prefers the chowder; his sister Eve prefers the other. Their preferences conflict. But it is acknowledged between them that Adam finds tomatoes really repulsive, and loves clams, whereas Eve can take clam chowder or leave it alone, but is moderately fond of tomato soup. They agree to have the chowder.

The children are convinced that Adam's preference for clam over tomato exceeds Eve's preference for tomato over clam. Are they right? I think so. I also think that you are not in a position to have an opinion, not having been present at the interaction, and not knowing the children. But I can tell you what makes me think that they have accurately compared the intensities of their preferences; and I expect that when you have heard my reasons, you will go along with my conclusion.

One thing you don't know is whether or not Adam is simply being contrary: whether Adam's expressed loathing for tomatoes and his expressed love for clams are expressions not of his taste in food but of his wish to frustrate his sister. You don't know, but I do, and so does his mother, and his sister. We can assure you that Adam is a genuine clam fan, and that he is genuinely disgusted by tomatoes. (Not as much, I would say, as *I* am disgusted by boiled tripe, but that's not at issue.) We have seen Adam in restaurants unhesitatingly choosing clams, when available, and anxiously inquiring into the

First published by R. Jeffrey, in *Logical Theory and Semantical Analysis*, S. Stenlund, ed., 1974. Reprinted by permission of Kluwer Academic Publishers.

tomato content of unfamiliar dishes when the choice is among them. And eschewing New York–style clam chowder, which contains tomatoes. We have also seen Eve choose clam chowder sometimes, and tomato soup more frequently. We also know that Eve can be as vehement as Adam in her attitudes toward other foods, and that her general style is more vivid than Adam's. It is unlike her, in general, to dissimulate her feelings and (given everybody's mood at this lunch) very difficult to imagine that the present issue is an exception. Nor does one of them generally manage to dominate the other. In particular, their agreement that Adam shall have his way at this lunch seems to none of us the expression of Eve's (nonexistent) submissiveness to her big brother.

It seems to me – and I think it should seem to you, now that you have heard the story – that Adam and Eve correctly agreed that the fair thing to do would be for them to have the chowder. Mind you, some of us, if pressed, might have expressed some reservations about the *quality* of Adam's loathing for tomatoes. Perhaps it's not the taste, and not some threatening subconscious idea about tomatoes in particular that puts him off them. Perhaps tomatoes are simply an item about which he successfully dug his heels in, on some forgotten occasion years ago when he was in a contrary mood and tomatoes were what we were trying to get him to eat. Perhaps he cherishes that victory, and even now uses tomatoes to reaffirm his autonomy, and test our acceptance of it. That's as may be. But to deny that his attitude toward tomatoes is radically gastronomical is not to deny its strength.

I have gone into so much detail in order to remind you of the complexity and variety of considerations which we bring to bear, in forming judgments about comparative intensities of preferences. Note well that some of the relevant data are extrapreferential, and that the theory of preference is not the only relevant theory. Thus, it is relevant in my example that Eve's style is generally more vivid than Adam's. That was a factual claim, couched in the language of a commonplace theory (or prototheory) in which people (anyway, some people) are said to have styles which are describable and comparable in various ways, e.g., in terms of vividness. Of course, this talk of comparative vividness implicates interpersonal comparisons of preferences: To a first approximation, the claim that Eve's style is generally more vivid than Adam's involves a claim that when their preferences are equally intense, Eve can be expected to use stronger language and more urgent gesture than Adam to express them.

Have I then *smuggled in* an assumption about interpersonal comparability of preferences in comparing different people's styles? Well, if it was smuggling, I've just confessed it, and unabashedly. It seems to me that preferences belong to a large family of attitudes and states which we impute to ourselves and others in ways which make it apt to compare intensities interpersonally as well as intrapersonally. Examples are *needs*, e.g., for food, which we share with plants and other animals; *pleasure* in activities like feeding and generating, which we share with other animals; and various distinctively human states and attitudes, e.g., Left Hegelianism, and the desire for a job in the East. These states and attitudes can implicate one another in familiar if hazy ways.

II. BUT PREFERENCE IS A TECHNICAL CONCEPT

This is all very well (you may say), but preference is a technical concept which, I have suggested, lives and moves and has its being in theories like those of Ramsey, and von Neumann and Morgenstern. In those theories the concept of preference is implicated, not with needs and pleasures and desires for jobs, but with the technical concept of judgmental probability and the more homely concept of choice or voluntary action. It is through this latter concept that preference (and through it, personal utility) is thought to acquire its character of observability or testability: its operationalistic legitimacy. But matters are not so simple, I think. Offered *A* or *B* as he will, the subject chooses *A*. This voluntary action evidences his preference for *A* over *B* or, more accurately, his nonpreference for *B* over *A*. But to function as evidence in this way, the event in question must genuinely be an act; must be voluntary; and must be deliberate: considered, not rash, not perverse. The choice, in other words, must have been a preferential choice. The question, whether a response was a preferential choice, may arise when we try to understand a sequence of events which seem to imply intransitivity of the subject's preferences. Sometimes, we use data about someone's preferences in order to determine the status of an event: whether or not it was a preferential choice of *A* over *B*. (We may

184

conclude that he misspoke himself, or that he was confused about which was which, or that he was trying to mislead us about his true preferences.) I am far from subscribing to the neopositivistic view which is adopted as a matter of course by many economists, psychologists, and others. I take it that in applying theories like those of Ramsey and von Neumann and Morgenstern we rightly connect such technical terms as preference and judgmental probability with homelier talk of hunger, doubt, and so on.

Still, I think it important to cite fairly clear theoretical contexts in which interpersonal comparison of preferences is important – if only because such intercomparisons have become a watchword among economists: a paradigm of hocus-pocus. To that end, I centered Essay 9, above, around the rather technical work of the economists Fleming and Harsanyi – work which demonstrates that utilitarianism has great prima facie plausibility as a norm for social decision-making, and which, turned wrong way around, provides an interpersonal analogue of the Ramsey–von Neumann–Morgenstern account of *intra*personal comparison of preference intensities. It turns out that if social preferences, like personal preferences, are to satisfy the von Neumann–Morgenstern axioms, and if the social preferences are related to those of the individuals in two very plausible ways, then it is always possible to view the social preferences as having been obtained from those of the individuals in a straightforwardly utilitarian manner: There will be a utility function which represents the social preferences, and utility functions for the various individual preferences which yield the social utility function by straightforward Benthamite summation. Furthermore, all of these utility functions will be determined by the various preference rankings as uniquely as need be, for purposes of interpersonal comparison of preference intensities! The two plausible conditions are *functionality* and *positivity,* (14) and (15) on p. 177, above.

I take this to be a way of connecting the notion of interpersonal comparison of preferences with the notion of even-handed compromise between individuals, e.g., Adam and Eve in my example. I take it that we can sometimes determine not only the preferences of the individuals but also the social preferences which they have arrived at, *and can also determine the fact that they regard those social preferences as representing an even-handed compromise between their conflicting personal preferences.* In such cases, the

185

Fleming–Harsanyi–Domotor results allow us to find commensurate unit intervals for the personal utility scales which are involved, i.e., we can perform interpersonal comparisons of preferences.

Let me reiterate that I do not therefore subscribe to utilitarianism as a general norm for social decision-making. But I do think that there are special circumstances where it is appropriate – fairly common circumstances.

III. COLLATION OF PREFERENCES

In case "society" consists of just two individuals – Adam and Eve, say – Harsanyi's results can be stated as in condition (20) of Essay 9 on p. 179, above: If Adam is indifferent between V and W and Eve is indifferent between X and Y, then the intensity of Adam's preference for X over Y is greater than or less than or the same as the intensity of Eve's preference for V over W accordingly as "society" prefers a 50–50 gamble between X and W to a 50–50 gamble between Y and V, or has the opposite preference, or is indifferent between the two gambles. In our example, where $X = W =$ clam and $Y = V =$ tomato, this scheme simplifies down to: Adam's preference for clam over tomato exceeds Eve's preference for tomato over clam because society prefers clam to tomato, i.e., prefers a 50–50 gamble between clam and clam to a 50–50 gamble between tomato and tomato. Then this is a particularly vivid illustration of how direction of social preference can reveal comparative intensity of individual preferences, when social and individual preferences have the characteristics which Harsanyi demonstrated to be necessary and sufficient for social decision-making to be viewable in a neo-Benthamite light.

But use of this criterion requires prior success in the project of forming a social preference ranking which represents an even-handed compromise between the conflicting personal preference rankings. Where such a social preference ranking is not available, however, other methods may be used, with the aid of which an even-handed social compromise can be identified, in favorable circumstances. Our example is a case in point, for I take it that the children compared the intensities of their preferences by *collating* the relevant parts of their personal preference rankings, as follows:

186

Adam's	*Eve's*
Preferences	*Preferences*
Clam	
	Tomato
	Clam
Tomato	

Thus, they agree that *Adam likes clam more than Eve likes tomato,* and *Eve likes clam more than Adam likes tomato:*

$$u_{Adam}(\text{clam}) > u_{Eve}(\text{tomato})$$
$$u_{Eve}(\text{clam}) > u_{Adam}(\text{tomato}).$$

Note that these are absolute intercomparisons of utilities. Coupling them with the unproblematical claim that Eve prefers tomato to clam,

$$u_{Eve}(\text{tomato}) > u_{Eve}(\text{clam}),$$

we have the result that Adam's preference for clam over tomato is more intense than Eve's preference for tomato over clam:

$$u_{Adam}(\text{clam}) - u_{Adam}(\text{tomato}) > u_{Eve}(\text{tomato}) - u_{Eve}(\text{clam}).$$

Thus we have our interval intercomparison of utilities, from which it follows that utilitarian society ought to prefer clam to tomato: Transposing, we have

$$u_{Adam}(\text{clam}) + u_{Eve}(\text{clam}) > u_{Adam}(\text{tomato}) + u_{Eve}(\text{tomato}).$$

The absolute intercomparisons of utilities which led to this interval intercomparison and neo-Benthamite compromise have their theoretical home in the scheme for social decision-making which I noticed briefly at the top of p. 651 of my *Journal* article. To apply that scheme one needs something more than the interval comparability of utility scales which suffices for the neo-Benthamite scheme. The idea behind the new scheme is that different individuals' likes and dislikes among social prospects should sometimes be weighted differently, in forming the social sum. If I know more about farming than you, my attitudes about agricultural policy may deserve more weight than yours in forming society's attitude. If you use libraries more than I do, your attitude about hours when the libraries shall be open may deserve more weight than mine. And if you represent a group whose interests have been systematically ignored in the past when questions of schooling and public housing were at issue, your attitudes on such matters may deserve more weight than mine.

IV. FACTUAL JUDGMENT OR MORAL IMPUTATION?

Schick[1] holds that the question, "What are Adam's personal preferences," like the corresponding question about Eve, is a factual one, but that it is pointless or perhaps meaningless to ask whether one of Adam's preferences is more intense than one of Eve's. Instead, he proposes a Procrustean solution: Adopt identical upper limits for their two utility scales, and common lower limits, and justify this imputation on moral grounds, viz., that in that way we are treating them alike. Having imposed absolute commensurability in this way, we obtain an *assimilation of* their utility scales which he declines to describe as a comparison.

Schick says,

Some people are said to be capable of greater intensities of feeling than others. The meaning of this is in doubt, and so of course also its truth. But however the claim is understood, I do not see why it should concern us. Adam values his *summum bonum* as highly as he values anything, and his *summum malum* is for him the worst of all possibilities. The same is true for Eve. Why then should Adam's voice on his *extrema* be given any weight different from that given Eve's voice on hers? . . . why should a fanatic count for more than a person with tired blood? I see no reason why he should, and so have equalized the limits of the utility ranges. (665-65)

As you will have suspected, I find this unconvincing. First, interpersonal comparison of preference intensities is not a matter of interpersonal comparison of intensities of *feeling*, any more than intrapersonal comparisons are. To speak of feelings here is to invite specious questions: "How can I feel your pains?" and the like. But Eve's judgment about Adam's preference for clam over tomato is no pretense to interlocking of nervous systems, or sharing of sense data. It is based on empathy, if you like, but that empathy is no occult intuition. Rather, it is an *attitude* with which she observes and recalls his behavior at table, in our kitchen and in restaurants. That attitude is the one we normally adopt in observing and interpreting the behavior of other people. It is a matter of viewing

1. "Beyond Utilitarianism," *J. Phil.* **68** (1971), 657–666.

them and treating them *as people*. What constitutes treating someone as a person may vary significantly from culture to culture and
from era to era within a developing culture; and at various stages of
various cultures it may be thought right to treat cows or dogs or
other animals as people, and to decline to treat certain humans as
people. Then the scope and perquisites of the status, *person*, are
time- and culture-relative; to hold that the status belongs to every
human is to make a moral commitment; and to judge that Adam's
preference for clam over tomato exceeds Eve's preference for tomato over clam is to go beyond bare facts (if bare facts there be).

I suppose that Schick may go along with this; but he goes further.
He suggests that, implicit in the commitment to treat humans as
people is a commitment to treat everybody's highest goods on a par,
and similarly for the *infima mala* at the bottoms of their preference
rankings. But I see no reason to believe that; indeed, I think we need
not go far to fetch cases in which adoption of the attitude that Adam
and Eve are people implies (given the facts of the particular cases)
that their preferential *extrema* should not be treated on a par. Surely
the argument is not advanced by the tautologous observations that
Adam values his *summum bonum* as highly as he values anything,
that his *infimum malum* is for him the worst of all possibilities, and
that the same is true for Eve. Surely the question remains, whether
Adam's *summum bonum* merits the same weight and the same utility
as Eve's, when it is time to make a social decision. What would we
say in a case where an older Adam's *summum bonum* is identical
with Eve's *infimim malum*, viz., Eve's imminent death, whereupon
Adam will set up house with Lillith?

Why should a fanatic count for more than a person with tired
blood? Surely there is no *general* reason why he should, but as
surely there are cases where he should, e.g, *some* cases where the
"fanatic" is an Eve in the strength of youth, with her life before her;
where the person with "tired blood" is an Adam dying of leukemia,
with a week to live; and where the consequences of their social
decision will be faced next year by Eve, but not by Adam.

These two examples need further study (which I shall not undertake here), perhaps in the light of the nonegalitarian aggregation
schemes (6) and (7) of Essay 9 above. There are threads in these
examples which ought to be separated, but they are the same threads
which, I think, need separation in Schick's argument for his assim-

189

ilation scheme, viz., assumptions about aggregation and assumptions about intercomparison or assimilation.

Let me conclude by returning to the issue of clam versus tomato. I grant the prima facie attractiveness of hypothesis that Adam values his *summum bonum* precisely as much as Eve values hers, and similarly for their *infima mala;* and I note that clam is not Adam's *summum bonum,* nor is tomato Eve's. (Rather, those items are *extrema* in tiny fragments of their overall preference rankings.) But there is another hypothesis which has its own strong attractions in the light of the facts about Adam and Eve as I have recounted them; and this second hypothesis may well conflict with Schick's in the light of global characteristics of the two personal preference rankings, outside the small fragments we have been discussing. To repeat: Schick's hypothesis (or, his method of assimilation) determines a definite collation of Adam's and Eve's total preference rankings, and that collation may conflict with the collation arrived at above, of the clam and tomato fragments. In effect, Schick proposes that any such conflict be resolved in favor of the collation which follows from his assimilation scheme. But what if we consider a variety of small fragments of the two personal rankings, corresponding to various social decision problems, and find plausible collations in each case which, viewed globally, prove to be consistent with each other but not with Schick's uniform collation? I suggest that in such a case one would abandon Schick's hypothesis, despite its abstract, prima facie attractions.

But let me not overemphasize my differences with Schick on the matter of intercomparison. (His general treatment of dependencies, which I find very attractive, is compatible with other modes of assimilation.) I may be wrong; and it would be no bad thing if Schick were right, and the interpersonal comparisons (or assimilations) of preferences obtained by mapping everybody's preferences onto the unit interval can be relied upon to agree with substantial bodies of local collations and weaker local intercomparisons of preference intervals. Nor, I suspect, do we differ much about what constitutes being right, here: We agree that the question is (at least in large part) a moral one, and we agree that standardization on the unit interval has strong attractions as a moral principle, viz., simplicity and universality. My contention is that there are cases fairly close to home in which Schick's principle looks not simple but

190

simplistic, and in which its generality looks like Procrustean indifference to the facts and to the moral perquisites of the status, *person*. In support of this contention I have cited three cases (*lunch, Lillith,* and *leukemia*) and two methods (interval intercomparison after Harsanyi et al. and absolute intercomparison via collation) which, where applicable, can yield interpersonal comparisons or assimilations in which the utility differences between the *extrema* of different people's preference rankings cannot be equal.

11

Mises redux

Once one has clarified the concept of random sequence, one can define the probability of an event as the limit of the relative frequency with which this event occurs in the random sequence. This concept of probability then has a well defined physical interpretation. (Schnorr, 1971, pp. 8–9)

Mises' (1919) concept of *irregular* ("random") *sequence* resisted precise mathematical definition for over four decades. (See Martin-Löf, 1970, for some details.) This circumstance led many to see the difficulty of defining "irregular" as *the* obstacle to success of Mises' program, and to suppose that the solution of that difficulty in recent years has finally set probability theory on the sure path of a science along lines that Mises had envisaged. To the contrary, I shall argue that since stochastic processes do not go on forever, Mises' identification of each such process with *the infinite sequence of outputs it would produce if it ran forever* is a metaphysical conceit that provides no physical interpretation of probability.

1. BERNOULLI TRIALS

Martin-Löf (1966) showed how to overcome the distracting technical obstacle to Mises' program, and Schnorr (1971) and others have continued his work. The air is clear for examination of the substantive claim that probabilities can be interpreted in physical terms as limiting relative frequencies of attributes in particular infinite sequences of events.

The simplest examples are provided by binary stochastic processes such as coin-tossing. Here, Mises conceives of an unknown member, *h*, of the set of all functions from the positive integers to the set {0, 1} as representing *the* sequence of outputs that the process

First published by R. Jeffrey, in *Basic Problems in Methodology and Linguistics*, R. E. Butts and J. Hintikka, eds., 1977. Reprinted by permission of Kluwer Academic Publishers.

would produce if it ran forever. He then identifies the physical probabilities of attributes as the limiting relative frequencies of those attributes in that sequence; e.g., in the case of tosses of a particular coin, h is defined by the condition

(1) $h(i) = 1$ *iff the ith toss (if there were one) would*
                    yield a head,

and the probability of the attribute *head* is defined,

(2) $$p(head) = \lim_{n \to \infty} \frac{1}{n} \sum_{i=1}^{n} h(i).$$

Both parts of this definition are essential to Mises' attempt to interpret $p(head)$ as a physical magnitude.

In their algorithmic theory of randomness for infinite sequences, Martin-Löf, Schnorr, et al. have provided satisfactory abstract models within which part (2) of the definition makes mathematical sense. Thus, Martin-Löf (1968) proposes a model in which Mises' irregular collectives are represented by the set of all functions h that belong to all sets of Lebesgue measure 1 that are definable in the constructive infinitary propositional calculus, e.g., the set of sequences for which $p(head) = 1/2$ in (2). In proving that the intersection of all such sets has measure 1, he shows that his definition escapes the fate of von Mises' (according to which there would be no random sequences) and yields the desired result, that "almost" all infinite binary sequences are random. The condition $p(head) = 1/2$ is inessential: The same approach works for Bernoulli trials with any probability of *head* on each.

But the brilliance of this abstract model of Bernoulli trials is far from showing how probability is connected with physical reality: Rather, it deepens the obscurity of Mises' condition (1), which purports to provide that connection. For most coins are never tossed, and those that are, are never tossed more than finite numbers of times. No infinite sequence of physical events determines the function h of (2): For all but a finite number of values of "i," the clause following "iff" in (1) must be taken quite seriously as a counterfactual conditional. But unless the coin has two heads or two tails (or the process is otherwise rigged), there is no telling whether the coin would have landed head up on a toss that never takes place. That's what probability is all about.

A coin is tossed 20 times in its entire career. Would it have landed

193

head up if it had been tossed once more? We tend to feel that there must be a truth of the matter, which could have been ascertained by performing a simple physical experiment, viz., toss the coin once more and see how it lands. But there is no truth of the matter if there is no 21st toss. The impression that there *is* a truth of the matter arises through the analogy between (a) extending a series of tosses of a coin, and (b) extending a series of measurements of a physical parameter, e.g., mass of a certain planet. If $p(head)$ is a physical parameter on a par with m(Neptune), then – the argument goes – (a) really is just like (b). But the analogy is a false one because while Neptune exists, and has a mass whether or not we measure it to a certain accuracy, the 21st toss of a coin that is tossed only 20 times does not exist and has no outcome: Neither *head* nor *tail*. A truer analogy would compare p(head) with $m(x)$ where x is a nonexistent planet, e.g., the 10th from the Sun. Mises defines $p(head)$ as the limiting relative frequency of heads in an infinite sequence that has no physical existence. If one could and did toss the coin forever (without changing its physical characteristics) one would have brought such a sequence into physical existence, just as one would have brought an extra planet into existence by suitable godlike feats, if one were capable of them and carried them out. But in the real world, neither the sequence nor the planet exists, and the one is as far from having a limiting relative frequency of heads as the other is from having a mass.

Granted: There is a telling difference between the two cases. In the case of the nonexistent 10th planet we are at a loss to say what its mass would be if it had one, while in the case of the coin that is tossed just 20 times we are ready enough to name a probability for heads. If the coin is a short cylinder with differently marked ends and homogeneous mass distribution, we are confident that heads have probability $1/2$. But this difference tells against Mises: It identifies the probability of heads as a physical parameter of the coin, whether or not it is ever tossed, in terms of which we explain and predict actual finite sequences of events – directly, and not by reference to a nonexistent infinite sequence of tosses. It is because the probability of heads is $1/2$ that we grant: If the coin *were* tossed ad infinitum without changing its physical characteristics, the limiting relative frequency of heads would be $1/2$. But since there is no infinite sequence of tosses, "its" characteristics cannot explain why heads have probability 1/2.

2. IRREGULAR FINITE SEQUENCES

In the 1960s, Kolmogorov and others (Chaitin, Solomonoff) founded a theory of algorithmic complexity of finite sequences that sheds fresh light on probability. In showing that the sequences irregular in Kolmogorov's sense are those that pass a certain universal test for randomness, Martin-Löf (1966) provided an alternative definition of irregularity that he was able to extend quite naturally to the case of infinite sequences. In deprecating the foundational importance of the infinite case, I am far from denying the importance and foundational relevance of the finite case as treated by Kolmogorov, Martin-Löf, and others. What I do wish to deny is that by continuity with the finite case, or by mathematical infection from it, the infinite case gets the importance it would have if ours were a world in which each Bernoulli process went on forever (and in which each Markov process, infinitely replicated, went on forever). To get a sense of the importance and autonomy of the finite case, let us review it briefly.

Tables of "random numbers" are long, irregular sequences of digits – binary digits, let us suppose. The easiest and surest way to generate such sequences is by Bernoulli processes with equiprobability for the two outcomes on each trial, e.g., by repeated tosses of a coin, with heads recorded as 1's and tails as 0's. In principle, such a process could yield a table of a million 1's, but in practice, no one would buy such a table or give it shelf space.

Why? Well, why spend the money? The table is utterly regular, the relevant rule being, "Write 1 000 000 ones." It is not only cheaper but easier to use that rule in your head than to buy and consult the table. *Moral:* We use "equi-Bernoulli" processes to generate tables of "random numbers" not because we have use for the outputs of such processes no matter what they may prove to be, but because we expect such outputs to be irregular, and it is irregularity of the sequence that we seek, irrespective of its provenance.

Kolmogorov (1962) pointed to *incompressibility* as a definitive characteristic of irregularity of finite sequences. Thus, a string of 1 000 000 1's is compressible, for the rule "Write 1 000 000 ones"

195

would be only some 100 binary digits long if letters, digits, and spaces were coded in some fairly simple way as blocks of binary digits. In detail, questions of compressibility are relative to (1) choice of one out of the infinity of universal systems of algorithms or programming schemes for generating binary sequences, and to (2) choice of one out of the infinity of measures of complexity of algorithms belonging to the same universal system – let us say, via length of representation in one of the infinity of effective binary coding schemes. Once these choices have been made, we have the means to define the *irregular* ("random") finite sequences as those *about as long as the shortest binary coded algorithms that generate them*. If we define the *algorithmic complexity* of a finite binary sequence as *the length of the shortest binary coded algorithms that generate it,* then the irregular sequences are those whose lengths are approximately equal to their algorithmic complexities.

Locally (i.e., for each particular sequence) the relativity of algorithmic complexity to choices (1) and (2) is problematical [cf. Goodman's (1955) "grue" paradox], but globally its effect is negligible, for if k_1 and k_2 are two particular measures of algorithmic complexity, there will be a finite bound on the absolute differences between $k_1(s)$ and $k_2(s)$ as "s" ranges over all finite binary sequences. Thus, one proves that the percentage of irregular sequences among all sequences of the same length approaches 100 as the length of the sequences increases without bound: *For large n, practically all sequences of length n are irregular.*

Why do we turn to equi-Bernoulli processes as sources of irregular finite sequences? Kolmogorov's theory provides a clear answer, as follows. (1) For such processes, all output sequences of length n have probability 2^{-n}. (2) For large n, practically all sequences of length n are irregular. Therefore: (3) For large n, the probability is practically 1 that the output of such a process will be irregular. Then devices lie ready to hand that, with practical certainty, generate long irregular sequences. But mathematical certainty about irregularity is far more difficult to attain: (4) For universal systems of algorithms, the halting problem is unsolvable, and therefore there is no effective test for irregularity of finite sequences. In principle, one might nevertheless be able to prove that particular finite sequences are irregular, but in practice we do well to rest content with high probability.

3. MIXED BAYESIANISM

Suppose that a coin is tossed 40 times, and the process yields nothing but heads. There is a dim argument to the effect that this should not surprise us, for the sequence of 40 heads is no less probable than any other sequence of that length, be it ever so irregular. Of course, this argument must be wrong if we rightly see in such an output compelling evidence that the source was not as we had supposed it to be, e.g., if we see the output as overwhelming evidence that the source, far from being equi-Bernoullian, is one that yields heads with probability 1 on each toss. But what is the rationale behind this sensible view of the evidence? Here I give a mixed Bayesian answer to this question – "mixed" in the sense that while statistical hypotheses about the source are treated objectivistically (as hypotheses about physical magnitudes), probabilities of those hypotheses are treated judgmentalistically ("subjectivistically").

(1) Consciously or not, we do or should entertain various hypotheses $H_1, H_2, \ldots$ about the source, where (in the present example) initially we judge H_1 (the equi-Bernoullian hypothesis) to be overwhelmingly more probable than H_2 (the hypothesis that heads have probability 1 on each toss), in some such sense as this:

$$\frac{p(H_1)}{p(H_2)} = 2^{20} \approx 1\ 000\ 000.$$

(2) After seeing the output sequence and so verifying the evidence-statement $E =$ "The output is a string of 40 heads," we revise our judgment via the probability calculus, changing our degree of belief in each hypothesis H from its prior value, $p(H)$, to its posterior value,

$$p(H \mid E) = p(E \mid H) \times \frac{p(H)}{p(E)} \quad \text{(Bayes' Theorem)}.$$

so that now H_1 is overwhelmingly less probable than H_2, in the sense that

$$\frac{p(H_1 \mid E)}{p(H_2 \mid E)} = \frac{p(E \mid H_1)p(H_1)}{p(E \mid H_2)p(H_2)} = 2^{20}2^{-40} \approx 0.000\ 001.$$

In this Bayesian answer, the probabilities of the hypotheses are "subjective" in the sense that they are degrees of belief, which need

197

not be "subjective" in the sense of being ill-founded, arbitrary, or idiosyncratic. But the statistical hypotheses H_1 and H_2 themselves are treated objectivistically. Some Bayesians – notably, de Finetti – would treat all probabilities as degrees of belief, and others would treat all of them objectivistically. The mixed position represented here is the commonsensical version of Bayesianism that Bayesian extremists must explain away or reproduce within their own terms of reference.

"Bayesians" are so called because of their willingness to use Bayes' theorem in cases where most thoroughgoing objectivists would reject as senseless the prior probabilities $p(H)$ and $p(E)$ of evidence and hypothesis that appear in it. The affinity of Bayesianism with "subjectivism" (judgmentalism) derives from the fact that we may have broadly shared judgments in the form of degrees of belief in H and E even in cases like the present example, where prior inspection of the coin is supposed to have led us to think the equi-Bernoullian hypothesis overwhelmingly more probable than the other, but where we envisage no definite stochastic process of which the coin is the product – a process of which the ratio of *physical* probabilities would be $p(H_1)/p(H_2) \approx 1\ 000\ 000$. Pure objectivists who would be Bayesian must envisage some such higher-level process, and treat the prior probability function p as the probability law of that process. Thus, commonsense objectivists sometimes speak (without conviction) of urns containing assortments of coins, some normal, some bent, some two-headed, etc., out of one of which the coin actually used is imagined to have been drawn. In that vision, $p(H_1)$ is the proportion of normal coins in the urn.

Observe that where the "subject" thinks she knows the objective probability of an event (e.g., the event that all 40 tosses yield heads) and thinks she knows nothing else that bears on the matter, (e.g., perhaps, that the first toss yielded a tail!), she will adopt what she takes to be the objective probability as her degree of belief. Then "subjective" does not mean *whimsical*. To call a probability "subjective" is simply to say that it is somebody's degree of belief. One does not thereby deny that the belief has a sound objective basis. Furthermore, the "events" to which subjective probabilities can be attributed need have no special character (e.g., "unique," weird, etc.), for they are simply the events concerning which people can

198

have degrees of belief, viz., all events whatever. [These remarks are directed in part to the comments on subjective probability in Schnorr (1971, p. 10).]

4. NONFREQUENTIST OBJECTIVISM

Frequencies are important: The laws of large numbers tell us why; e.g., they tell us that in stationary binary processes, the relative frequency of "success" will in all probability be very close to the probability of success on the separate trials. Notice that here, the notion of probability appears along with that of relative frequency in the formulation of the law itself. (The notion of probability appears as well in the definition of "stationary," viz., invariance of probability of specified outcomes on specified trials, under translation of trials.) Frequentism is a doomed attempt to define probability in such a way as to turn the laws of large numbers into tautologies.

The lure of von Mises' program lies in its goal of providing a uniform, general definition of probability as a physical parameter – a definition that can be applied prior to the scientific discoveries that reveal the detailed physical determinants of stochastic processes, as e.g., the discoveries by Mendel, Crick, and many others revealing the mechanisms underlying the mass phenomena encountered in genetics. Mises sought to found probability as an independent science, on the basis of imaginary infinite sequences of events. Taxed with the unreality of those foundations, he replied that they are as real as the foundations of physics: To measure the physical parameter prob(*head*) to a desired accuracy it suffices to toss the coin often enough, for prob(*head*) is the limit of such a sequence of measurements just as surely as m(Neptune) is the limit of another sequence of measurements. Shall we hold the foundations of probability to a higher standard of physical reality than that to which we hold physics itself?

Surely not; but here, Mises holds physics itself to a remarkably low standard of reality, i.e., essentially, the idealist standard to which Bishop Berkeley held it: *Esse percipi est*. The suggestion is that the mass of Neptune exists to the extent to which we measure it, just as the sequence of outcomes exists to the extent to which we toss the coin. As was suggested in Section 1, the limiting relative frequency of heads in the "ideal" (i.e., nonexistent) infinite sequence of tosses is more properly compared with the mass of some nonexistent planet, e.g., the 10th from the Sun.

But if probabilities are not limiting relative frequencies, what are they? If there is no uniform, general definition of probability that is independent of other scientific inquiries, how shall we define probability as an objective magnitude? I would answer these questions as follows.

The physical determinants of probabilities will vary from class to class of cases; there is no telling a priori what they will prove to be. In the case of die-casting, the experience of gamblers and tricksters joins with physical and physiological theory to point to the shape, mass distribution, and (most important) markings of the die itself, as the determinants of the probabilities of the possible outcomes on each toss, and these considerations also join to say that different tosses are probabilistically independent. The case is similar for coin-tossing (where the point about markings is that there *are* two-headed coins about). In lotteries, by design, the determinants are the numbers of tickets of each sort (or the numbers of balls of different colors in the urn), but design is not enough: Empirical and theoretical inquiry may show the design to have been defective, e.g., because the balls of one color share a palpably distinct texture. As with games of chance, so with social, biological, and physical probabilities, but even more so: We look to experience, informed with theory, to identify the objective determinants of the probability laws of types of stochastic processes.

The easiest cases are lotteries and urn processes. There, we identify objective statistical hypotheses with the makeup of (say) the urn, and, by a happy accident, the probability of drawing a ball of a certain color is numerically equal to the proportion of balls of that color in the urn. In practically all other cases, such a numerical coincidence is lacking. The "classical" view tried to generalize that coincidence to all stochastic processes. The frequentist view tries to generalize a different coincidence – one that is probable where the law of large numbers holds. On a nonfrequentist objectivistic view, one must face the fact that typically, no such coincidence will be forthcoming – not uniformly in all cases, and not even differentially, on a case-by-case basis. Still, we are often in a position where we can be fairly sure that the relevant determinants, difficult as they may be to describe explicitly and in detail, are the same in two processes, as when we ascertain that two coins were cast in the same mold under similar conditions: Believing that the determinants are shape, mass distribution, and markings, and having good reason to think that these determinants were determined in the same way for the two coins, we have good reason to think that the same probability law will govern the two processes of tossing them – even though we are at a loss to specify the common shape or the common mass distribution except ostensively.

No pure objectivist, I think it important to use judgmental probabilities, e.g., as illustrated in §3 (in an extreme, simplified example). The present suggestion is that the objective statistical hypotheses to which judgmental probabilities are attributed in such cases will be hypotheses about various kinds of physical magnitudes, which we shall seldom be in a position to specify explicitly and in detail, but which we can often identify ostensively, well enough for our purposes, once we understand what the *kinds* of magnitudes are that determine the process at hand – kinds like shape, mass distribution, and marking.

This is a far cry from Mises' uniform, general identification of probability with a particular physical magnitude, found in all cases; but that magnitude does not exist.

REFERENCES

Chaitin, G. J.: 1966, "On the Length of Programs for Computing Finite Binary Sequences," *J. Assn. Computing Machinery* **13**, 547–569.
Chaitin, G. J.: 1969, "On the Length of Programs for Computing Finite Binary Sequences: Statistical Considerations," *J. Assn. Computing Machinery* **16**, 145–159.
DeFinetti, B.: 1974, 1975, *Theory of Probability*, Wiley, 2 vols.
Goodman, N.: 1955, *Fact, Fiction, and Forecast*, Harvard.
Kolmogorov, A. N.: 1963, "On Tables of Random Numbers," *Sankhyā*, Ser. A **25**, 369–376.
Kolmogorov, A. N.: 1965, "Three Approaches to Definition of the Concept of Information Content" (Russian), *Probl. Peredači Inform.* **1**, 3–11.
Martin-Löf, P.: 1966, "The Definition of Random Sequences," *Information and Control* **6**, 602–619.
Martin-Löf, P.: 1970, "On the Notion of Randomness" in A. Kino et al. (eds.), *Intuitionism and Proof Theory* (Proc. of Summer Conf., Buffalo, N.Y., 1968), North-Holland.

Mises, R. v.: 1919, "Grundlagen der Wahrscheinlichkeitstheorie," *Math. Z.* **5**, 52–99.

Mises, R. v.: 1928, 1951, *Probability, Statistics, and Truth* (2nd revised English ed.), Macmillan, 1957.

Mises, R. v.: 1964, *Mathematical Theory of Probability and Statistics.* Academic Press.

Schnorr, C. P.: 1971, *Zufälligkeit und Wahrscheinlichkeit*, Springer Lecture Notes in Mathematics **218**.

Solomonoff, R. J.: 1964, "A Formal Theory of Inductive Inference," *Information and Control* **7**, 1–22.

202

12

Statistical explanation vs. statistical inference

Hempel is not the first philosopher to have held that causal explanations are deductive inferences of a special sort. In the *Posterior Analytics*[1] Aristotle distinguishes a special sort of deductive inference – the demonstrative syllogism – in these terms:

By demonstration I mean a syllogism productive of scientific knowledge, a syllogism, that is, the grasp of which is *eo ipso* such knowledge.

He then lays down defining conditions for this special sort of inference:

. . . the premisses of demonstrated knowledge must be true, primary, immediate, better known than and prior to the conclusion, which is further related to them as effect to cause.

And he remarks,

Syllogism there may indeed be without these conditions, but such syllogism, not being productive of scientific knowledge, will not be demonstration.

Now we can fault this account on various grounds, but so can we fault contemporary accounts. We must give the old man credit; as he says at the end of the *Organon* (at the end of *De Sophisticis Elenchis*), his was the first book on logic; and he concludes,

. . . there must remain for all of you, or for our students, the task of extending us your pardon for the shortcomings of the inquiry, and for the discoveries thereof your warm thanks.

The affinities between the Hempelian and Aristotelian accounts of explanation may be obscured by differences in terminology. Thus, Aristotle speaks of syllogism, Hempel of deductive inference; and Aristotle speaks of knowledge, Hempel of explanation. But remember that "syllogism" was Aristotle's general term for deductive

First published by R. Jeffrey, in *Essays in Honor of Carl G. Hempel*, N. Rescher et al., ed., 1969. Reprinted by permission of Kluwer Academic Publishers.

1. Book 1, ch. 2. All citations from this work are from the Oxford translation.

inference – and do not tax him with what we now know to be his overly narrow view of the forms that such inferences can take. As to knowledge versus explanation, Aristotle says

We suppose ourselves to have unqualified scientific knowledge of a thing, as opposed to knowing it in the accidental way in which the sophist knows, when we think that we know the cause on which the fact depends. . . .

It is precisely this sort of understanding that is conveyed by causal explanation – "the grasp of which is *eo ipso* such knowledge." Then I take it that Aristotle's *demonstrated knowledge* – knowledge, as he says later, not simply of the fact but of the *reasoned fact* – is the psychological correlate of Hempel's causal explanation as it is of Aristotle's demonstrative syllogism.

Of course, I have been overemphasizing the similarities between the Aristotelian and Hempelian accounts of scientific understanding; and of course, my point is not to advocate a back-to-Aristotle movement. Still less do I seek to belittle Hempel's work on explanation; the point is rather to appreciate that work as continuous with the philosophical enterprise that started in Athens 2400 years ago.

My real concern here is not with the history of nomological-deductive accounts of explanation – interesting as that may be – nor is it with the nomological-deductive view of causal explanation itself – satisfactory as *that* may be. My concern is with statistical explanation, and I have begun by recalling a general feature of nomological-deductive explanation because I want to show that statistical explanations often lack that feature. The particular feature is that of *being an inference*. On Aristotle's account as on Hempel's, causal explanations are deductive inferences; and it seems plausible to say that for their part, statistical explanations are statistical inferences – or, if you prefer, inductive or probabilistic inferences. But I shall argue that only in certain cases is it plausible to view statistical explanations as statistical inferences.[2]

The really beautiful cases of statistical explanation are the statistical mechanical ones in which the explained occurrence can be shown to have probability so close to 1 as to "make no odds" in any gamble or other decision problem. I am thinking here of such cases as the explanation of why a flat tire does not spontaneously inflate:

2. I leave to one side the deductive-statistical explanations discussed in part 3.2 of *Aspects of Scientific Explanation*, The Free Press, New York, 1965.

It is *possible* that the random movements of the surrounding air molecules might, for a few seconds, be such as to constitute a jet of air through the puncture powerful enough to pump the tire up – possible, but so improbable as to make it a *practical certainty* that no such jet will be forthcoming. (A proposition is a practical certainty if its probability is so high as to allow us to reason, in *any* decision problem, as if its probability were 1.) It is in these beautiful, extreme cases that the view of statistical explanations as statistical inferences is no less plausible than the view of causal explanations as deductive inferences.[3]

But there is a certain obliqueness about these inferential explanations, whether the strength of the inference be deductive certainty or practical certainty. I mean, one explains in order to impart knowledge of *how* or *why* the explained phenomenon takes place, but the explanation itself (in these cases) takes the form of a proof that the phenomenon *does* take place. The explanation is just the sort of thing one might produce in order to prove *that* the explained phenomenon takes place; this is the famous parallelism between explanation and prediction which I think breaks down for statistical explanations that impart less than practical certainty to the phenomenon explained.

I say that causal explanations (and the "beautiful" kind of statistical explanations) are oblique: They explain *how* or *why* by demonstrating *that* in a special way. (Compare Aristotle's conception of knowledge of the *reasoned* fact.) Of course, not every demonstration *that* serves to explain *why,* as Louis MacNeice's rhyme testifies:

The glass is falling hour by hour, the glass will fall forever.
But if you break the bloody glass you won't hold up the weather.[4]

The falling of the glass does not cause the weather to turn bad; it is rather an effect of something that the bad weather is also an effect

---

3. The notion of practical certainty operative in this account of the "beautiful" cases of statistical explanation is problematical but, I think, defensible. Example: There would be serious trouble with that notion if, given a probability greater than 0 (no matter how slightly greater), one could name a prize so great that the prospect of getting that prize with that probability is not negligible. But I take it that such problems as the St. Petersburg paradox already force us to realize that if Bayesian decision-theory is to work, there must be a finite upper bound on the utilities of the things the agent can envisage as prizes.

4. The lines from Louis MacNeice are the last two of his "Bagpipe Music."

of. Suppose it were true that *whenever* the glass falls, the weather turns bad. Then the inference

The glass is falling.
Whenever the glass falls the weather turns bad.
∴ The weather will turn bad.

would support a prediction *that* the weather will turn bad without explaining *why* the weather turns bad. On the other hand, the inference

The glass is falling.
Whenever the glass is falling the atmospheric pressure is falling.
Whenever the atmospheric pressure is falling the weather turns bad.
∴ The weather will turn bad.

would serve to both predict and explain the weather: It gives us knowledge not merely of the fact, but of the reasoned fact, for it proves that the fact is a fact by citing causes and not mere symptoms. In general, where the inference meets certain conditions, one of which is that a causal law appear among the premises, deductively grounded prediction will double as explanation. In these cases, knowledge *that* is imparted in such a way as to provide knowledge *why* as well. I mention this in order to suggest that this way of telling why obliquely, by demonstrating that, may not be the only way or the best way of telling why: the fact that we can sometimes explain by inferring is not a very strong reason to suppose that we can explain by *only* inferring. Perhaps, then, we should have another look at causal explanations, to see whether they *can* always be cast in the Aristotelian-Hempelian mold; but that is for another occasion. Here I want to have a look at statistical explanations, where I think it can pretty readily be seen that many of them simply cannot be represented as statistical inferences which show *why* obliquely, by demonstrating *that*.

Consider the following set of coin-tossing examples. A fair coin is tossed n times ($n = 1, 2, 3, \ldots$), and at least one head turns up. Here the Hempelian statistical inference would have as its premise a statistical law indicating that the probability of a head on any toss is $1/2$ and that distinct tosses are statistically independent, and would have as its conclusion the statement $H_1 \vee H_2 \vee \ldots \vee H_n$ that at least one of the first n tosses yields a head. The premise – the statistical law – might be specified in various equivalent ways, e.g., by saying that the probability of the conjunction of any k distinct H's is $1/2^k$. From this specification one can deduce, via the laws of the elementary probability calculus, the statistical probability of any truth functional compound of the H's, e.g., the statistical probability of the conclusion of the statistical inference which Hempel represents as follows:

$$\frac{\text{If } i_1, \ldots, i_k \text{ are all distinct, then } p(H_{i_t} \& \cdots \& H_{i_k}) = \frac{1}{2}^k}{H_1 \vee H_2 \vee \cdots \vee H_n} [1 - 2^{-n}]$$

The bracketed number at the right represents the inductive probability – the degree of confirmation – of the conclusion, conditionally upon the premise. If we represent the premise by "P" and the conclusion by "C" then the foregoing array is to be viewed as a typographical variant of the statement

$$c(C, P) = 1 - 2^{-n},$$

which gives $1 - 2^{-n}$ as the inductive probability or degree of confirmation of the conclusion on the premise.[5]

Hempel would hold that with $n = 1$ the foregoing inductive inference is too weak to serve as an explanation of the fact that there was at least one head on the first n tosses: with $n = 1$ the bracketed number – what Hempel calls *the strength of the explanation* – is only $1/2$. At the other extreme – for such large values of n as 100 – even I will admit that the inference will serve as an explanation, for a gap of $1/2^{100}$ between the probability of the conclusion and 1 is so small as to make no odds in any deliberation. But Hempel might hold that even with $n = 10$ we have an explanation: The bracketed "strength" of the inference is then $1023/1024$, or a bit over .999.

Now I agree that with $n = 10$ the strength of the inference is great enough to justify giving long odds, on the order of $1000:1$, on there being at least one head. The number $1023/1024$ is a good measure

5. I would prefer to avoid this second kind of probability and speak simply of the statistical probability of the conclusion; indeed, $p(C)$ is $1 - 2^{-n}$, i.e., the probability measure defined in the premise assigns to the conclusion precisely the value which the inductive probability measure c assigns to the conclusion conditionally on the premise. But let us stay with Hempel's way of talking.

of the proper strength of our expectation of the fact; but I deny that it is a good measure of the quality or strength of the explanation which the inference gives us. Indeed, I think it misleading to think of the statistical inference as being an explanation at all. The explanation how or why there was at least one head, I should think, would be given by specifying the statistical probability function p that governs the process, e.g., as in the premise of the inference, above. Perhaps it *is* part of the explanation to point out (as in the brackets above) that the probability of the explained phenomenon is then $1 - (1/2^n)$, *but the strength of the explanation would be no greater with $n = 10$ than with $n = 1$!* To say, in the case $n = 1$, that the statistical probability was $1/2$, is as much of an explanation as one can give. Indeed, this is a causal explanation in a sense in which a proof that there is at least one head can never be, for to say that the statistical probability is $1/2$ is to say *directly* something about cause which is said only obliquely, if at all, in a proof that there is at least one head. What is being said about cause, when we say that the statistical probability is $1 - (1/2^n)$ that there will be at least one head, is that this effect is "caused" by a chance or random process: We are saying directly that the usual sort of causality is absent.

To see this point it is helpful to consider what we should say if the improbable happened, say in the case $n = 2$, and there were two tails. The strength of the "explanation" of $-H_1$ & $-H_2$ would then be $1/4$ on Hempel's account, but we still have as complete an understanding of the why and the how of the outcome as we would have had if a head had turned up and the outcome had been the more probable one.

Moral. The strength of a statistical explanation (except in Hempel's technical sense) is not given by the degree of confirmation that the premises bestow on the conclusion in the corresponding Hempelian inductive inference. Similarly for $n = 10$: it is possible, although highly unlikely, that there will be ten tails, and if this happens we shall know all there is to know about the why of it and the how, when we know that the process which yielded the ten tails is a random one and when we know the probabilistic *law* governing the process. The knowledge that the process was random answers the question, "Why?" – the answer is, "By chance." Knowledge of the

208

probabilistic law governing the process answers the question "How" – the answer is, "Improbably, as a product of such-and-such a stochastic process." Note that knowledge of the probabilistic law of the process makes it at most a matter of calculation to find that the statistical probability of the phenomeonon – ten tails – is 1/1024. To know the probabilistic law is to know, among other things, that the actual outcome was hardly to have been expected.

Here knowledge *why* splits clearly away from knowledge *that*. Indeed in the statistical case I find it strained to speak of knowledge *why* the outcome is such-and-such. I could rather speak of *understanding the process* which had the outcome, for the explanation is basically the same no matter what the outcome: It consists of a statement that the process was a stochastic one, following such-and-such a law. (One may gloss this statement by pointing out that the actual outcome had such-and-such a probability, given the law of the process; but this gloss is not the heart of the explanation.)

Put the matter in this way: Aristotle's *knowledge of the reasoned fact* just will not do in the statistical case, for where a statistical explanation is appropriate, there's *no* reason for the fact: It came about by chance. Nor is the situation changed when (as usual) the probable happens. Because such cases are usual, we can usually give a statistical *inference* of strength 1/2 or more than we can give a statistical explanation; and this, I take it, is why it is easy to mistake the inference for an explanation. That the inference is *not* an explanation is shown, I think, by the fact that even when the improbable chances to happen, we give the same sort of account: The happening was the product of a *stochastic* process following such-and-such a probabilistic law. And we gloss this by pointing out that in this case the unexpected happened. My point is that it is no less a gloss, and no more essentially a part of the explanation, when we point out in the more usual cases that the *expected* happened.

Let us conclude by examining some cases in which explanation can be both causal and statistical. *Why was my first child a boy?* There are two sorts of answers: the statistical one ("There is no damned reason – it was pure chance"), and the causal ("Because the germ cell I contributed to the zygote which developed into the child was of the Y genotype"). Both answers are right; they supplement each other. The full-blown causal explanation might look like this:

The sperm which united with an ovum to form the zygote out of which my first child developed had the Y genotype.
Whenever the sperm which unites with an ovum to form the zygote out of which a child develops has the Y genotype, the child is a boy.

My first child was a boy.

Here the first premise could not have been known to be true before the birth, and so the inference could not have been used to support a prediction of the sex of the child; but it is a perfectly satisfactory causal explanation for all that. The second premise of the explanatory inference is a causal law. Its converse is also true, and can be used in conjunction with the conclusion of the explanatory inference to deduce the first premise of that inference. Indeed, that is how I know the genotype of the successful germ cell! But the deduction which tells me that is no part of the explanatory inference; it is rather part of the business of verifying that the explanatory inference really does meet the conditions for being a causal explanation. ("The premises of demonstrated knowledge must be true" among other things.) The causal and statistical explanations can be made to match by referring the statistics not to the sex of the child, but to the genotype of the relevant sperm cell: The usual process whereby spermatozoa unite with ova is a stochastic one in which the statistical probability is $1/2$ that the winner will have the Y genotype.

This is a case where before the outcome is known, we know that one of two causal explanations of the outcome will be correct – and we shall know *which* explanation is correct when we know the outcome. Before the event we had a lottery with causal explanations as prizes.

The same sort of thing can happen after the event, as in this made-up example: I draw at random from a box containing a two-headed penny and a normal one, toss, observe that a head turned up, and return the penny to the box without examining it further. Here the lottery is between a statistical and a causal explanation; the outcome of the lottery is not settled by the outcome of the toss, as it would have been if a tail had turned up; and we end in a position where as far as we know, the question "Why" may or may not have an answer!

Finally, consider the question of why my car would not start this morning. Here the question surely has an answer: There *is* a causal explanation although I do not know what it is. Suppose that it is only one morning in about twenty that my car will not start, and suppose that when it *does not* start, then nine times out of ten it is because of the defective frammas which I never had fixed. Here the statistical component generates an epistemic lottery in which the prizes are causes. There is an answer to the question "Why?" and the statistics give us a strong clue: It is very likely to be the frammas. Indeed, the probability that it is the frammas is .9, but there is no temptation here to speak of a statistical explanation of strength .9. Rather, the statistical component of the explanation goes like this: Nonstarting is the product of a stochastic process according to which the frammas goes bad with probability $(1/20)$ $(9/10)$ = .045 and something else goes bad with probability $(1/20)$ $(1/10)$ = .005, and in either case the car definitely is prevented from starting. Given this stochastic law, the probability of nonstarting is only .05, and this would be the strength of the Hempelian inferential explanation of the nonstarting. But the description of the stochastic process is better than the number .05 would indicate, as a statistical explanation: As far as it goes, it is perfect, and it goes as far as statistics will take us. What is needed to round it out is an item of causal knowledge: Was it the frammas?

CONCLUSION

Sometimes statistics enter our understanding of phenomena via causal lotteries. In such cases the full explanation would have both a statistical and a causal component; and it is doubtful whether Hempel would want to apply his statistical inferential model in such cases. Among the cases where he would apply his model, some are of the "beautiful" variety where the strength of the inference is so great as to allow us to speak of practical certainty in this sense: The probability of the phenomenon to be explained is so high, given the stochastic law governing the process that produces it, as to make no odds in any gamble or deliberation. These are the cases where the inferential model of explanation seems unexceptionable. But where the strength of the inference is more modest, I think it simply wrong to view the inference as an explanation, and to identify the strength

211

of the inference with the strength of the explanation. To explain the phenomenon that there was at least one head in two tosses of a coin, I would point out that the process is stochastic with probability 1/2 of head on each toss, and with different tosses independent of one another. I would give the same explanation if matters turned out differently: if, improbably, there had been no head on either toss. The difference between the two cases would lie entirely with the gloss: In the first case one would point out that the probable happened, while in the second, one would point out that the improbable happened. But the strength of the explanation would be the same in either case. Finally I point out that since the probable happens more often than not, we are usually able to provide a Hempelian inference when we are able to give a statistical explanation; and I suggest that this is what gives the view that statistical explanations are statistical inferences its specious plausibility.

13

New foundations for Bayesian decision theory

Aristotle makes some use of the ordinal notion of *utility* or *desirability* in his account of deliberation, e.g., in *De Anima* III, 11 (434a):

whether this or that shall be enacted is . . . a task requiring calculation; and there must be a single standard to measure by, for that is pursued which is *greater*.

But the role of probabilities in deliberation was not clearly seen until much later, e.g., by the authors of the *Port-Royal Logic* (1662; IV, 16):

to decide what one should do to obtain a good or avoid an evil, it is necessary to consider not only the good and the evil in themselves, but also the probability that they happen or not happen; and to view geometrically the proportion that all these things have together. . . .

This is essentially the "Bayesian" account, in which the agent performs an act of maximum ("expected")[1] utility; where the utility of an act is a weighted sum of the utilities of the consequences that it would have in the various possible contingencies, and the weights are the subjective probabilities of (degrees of belief in) those contingencies conditionally upon the act's being performed.

But to use the Bayesian account we need a way of discovering what probabilities the agent does or should attribute to the possible contingencies, and what desirabilities he does or should attribute to the possible consequences of his acts. The values that ought to be attributed often seem easier to obtain or discuss than those that are in fact attributed. Certainly regarding probabilities, the objectivistic relative frequency view has been the one most commonly held, and it has most often been held together with a reluctance to believe that

First published by R. Jeffrey, in *Logic, Methodology, and Philosophy of Science*, Y. Bar-Hillel, ed. Copyright 1965 by Elsevier Science Publishers.

1. As will appear below, the customary distinction between utility and expected utility plays no role in the present account of deliberation.

much sense or use can be made of de facto subjective probabilities.[2]

Degrees of belief can be deduced from the desirabilities attributed to various prospects; for, roughly speaking, the probability of E is the desirability of what the agent would be just willing to give up in order to be sure of getting something of desirability 1 in case E happens. But how do you measure desirabilities? According to von Neumann and Morgenstern [6] desirabilities are to be measured in terms of probabilities: To say that the desirabilities of A, B, and C are x, y, and z is to say that the agent is indifferent between a guarantee of B and a gamble between A and C with probability $(y - z)/(x - z)$ of getting A and probability $(x - y)/(x - z)$ of getting C.

It was Frank Ramsey [7] who, in 1926, first solved the problem of determining both subjective probability and desirability in terms of the agent's preferences between gambles. For present purposes his procedure may be described as follows. Consider two sets of propositions: the *consequences*, and the *events*, relative to the agent in question. Define a *conditioned event* as the conjunction of a consequence with an event. A *gamble* is a proposition of form

(1) $$[C_1, E, C_2]$$

where C_1 and C_2 are consequences and E is an event. (1) may be read,

(2) $$C_1 \text{ if } E, C_2 \text{ if not,}$$

where the two occurrences of "if" have more than truth-functional significance.

The desirability of such a gamble is a weighted sum of the desirabilities of the possible outcomes, in which the weights are the probabilities of winning and of losing.[3]

(3) $$U([C_1, E, C_2]) = U(C_1E)P(E) + U(C_2\bar{E})P(\bar{E}).$$

The fact that the "if"s have more than truth-functional significance in (2) is shown by formula (3), for if we read (1) as

(4) $$(E \supset C_1)(\bar{E} \supset C_2)$$

2. The situation has changed considerably in the past decade, largely through the work of Savage [8], which has also had the effect of calling the earlier work of Ramsey [7] and deFinetti [2] to the attention of statisticians. For further references, see the selected bibliography in [4].

3. I use juxtaposition, the wedge, the horseshoe, and the bar as signs of conjunction, disjunction, material implication, and denial; "T" for the necessary proposition; and "F" for the impossible proposition.

or, equivalently, as

$$EC_1 \vee \bar{E}C_2,$$

the utility of the gamble would be

$$\frac{U(EC_1)P(E)P(C_1 \mid E) + U(\bar{E}C_2)P(\bar{E})P(C_2 \mid \bar{E})}{P(E)P(C_1 \mid E) + P(\bar{E})P(C_2 \mid \bar{E})}$$

The difference between this expression and the right-hand side of (3) is just the difference between the case in which E and $\bar{E}$ are thought to guarantee C_1 and C_2, and the case in which the probabilities of C_1 and C_2, conditionally on E and $\bar{E}$, need not be 1.

Ramsey showed (*existence theorem*) that if the preference ranking of gambles satisfies certain conditions, there will be functions U and P, defined on the conditioned consequences and on the events, respectively, such that the preference ordering of gambles is identical with the numerical ordering of their expected utilities as computed by (3). He also showed (*uniqueness theorem*) that if two pairs of functions, (U_1, P_1) and (U_2, P_2) both meet the existence conditions, there will be a positive number a and a number b for which we have

(5) (a) $P_2 = P_1$
     (b) $U_2 = aU_1 + b.$

Finally (*equivalence theorem*) if the pair (U_1, P_1) meets the existence conditions and if a is positive, the pair (U_2, P_2) which is defined by (5) will also meet the existence conditions.

My object here is to outline a new theory of subjective probability and utility which is in certain respects simpler and, I think, more satisfactory than Ramsey's.[4] The new theory has certain peculiarities which I shall now list.

1. The theory is *unified* in the sense that probabilities and utilities are attributed to precisely the same objects, viz., to the members of a class of propositions which is closed under the finite truth functional operations (conjunction, disjunction, denial), but from which the impossible proposition has been deleted. In contrast, Ramsey attributes desirabilities to consequences, conditioned consequences, and gambles, but not to events. By Stone's representation theorem,

4. And Savage's [8]. For proofs and fuller expositions, see Bolker [1] and Jeffrey [5].

215

there is no loss of generality if we construe propositions as sets of *possible states of the world*, or *states*, for short.[5]

2. The theory is *noncausal* in the sense that neither [, ,] nor any other such causal notion is taken as primitive.[6] In the theory we can discuss the agent's preference ranking of truth-functional analogues of gambles – of propositions of form (4) – but not of the gambles (1) themselves. In the present theory, the elementary truth-functional operations of conjunction, disjunction, and denial must do the work of Ramsey's operation [, ,], to the extent to which that work is done at all.

3. The role of the linear transformation (5)(b) in Ramsey's theory is played by the *fractional linear transformation*

$$(6) \qquad\qquad U_2 = \frac{aU_1 + b}{cU_1 + d}$$

in the present theory.[7] The requirement that a be positive in (5)(b) is here replaced by the requirement that the determinant of the transformation be positive:

$$(7) \qquad\qquad ad - bc > 0.$$

For the equivalence theorem here we also have the requirements

(8)   (a) $cU_1(A) + d > 0$,
     (b) $cU_1(T) + d = 1$,

where condition (a) is asserted for all A for which $U_1(A)$ is defined (thus excluding $A = F =$ the empty set of states), and in (b), T is the set of all possible states, i.e., the necessary proposition.

4. Transformation (6) may *change probabilities:* When U_1 is transformed into U_2 by (6), P_1 is transformed into a probability measure P_2 which need not be identical with P_1. The overall situation is most simply described by defining

(9)   (a) $I_1 = U_1 P_1$,
     (b) $I_2 = U_2 P_2$,

5. The states will be the maximal sum ideals in the Boolean algebra of propositions, if propositions are initially construed in some other way than as sets. See Halmos [3], §18.
6. The argument following formula (3) above is designed to support the claim that the brackets represent a non-truth-functional operation which, in fact, I take to be a causal operation. This point is discussed further in [5].
7. This was first pointed out to me by Professor Gödel, who also sketched a statement and proof of the uniqueness theorem for the present theory. A somewhat different form of these results had been arrived at earlier (unknown to Gödel and to me) by Ethan Bolker [1].

216

so that I_n is a measure function, like P_n. In particular, I_n will vanish where P_n does, so that by the Radon–Nykodym theorem there will be an f_n such that

(10) $$I_n(A) = \int_A f_n \, dP_n.$$

Then by (9) we have

(11) $$U_n(A) = \frac{1}{P_n(A)} \int_A f_n \, dP_n:$$

The function f_n assigns a "utility" $f_n(s)$ to each state, s, and $U_n(A)$ is then the weighted average of the utilities of the states in which the proposition A is true, the weights being determined by the corresponding probability function, P_n. The transformation (6) can then be represented by a pair of transformations:

(12) (a) $I_2 = aI_1 + bP_1$,
(b) $P_2 = cI_1 + dP_1$.

Then by (9)(a) we have

(13) $$P_2 = P_1(cU_1 + d),$$

so that $P_1 = P_2$ if and only if $c = 0$, in which case d must be 1.

I take it that the first two of these peculiarities of the present theory are clear improvements over Ramsey's theory, but that the remaining two require fuller examination and defense.

The unity of the theory is as it should be. Propositions are commonly taken to be the objects of belief, but other sorts of concrete and abstract objects are normally spoken of as the objects of desire. But to desire a certain job, or someone's love, or a ham sandwich is to desire that one or another proposition hold: that the desirer *have* the job, or the love, or the sandwich, in an appropriate sense of "have."

As to the noncausal character of the present theory, notice first that the relationship between the agent's desirability function, U, and the corresponding probability function, P, is given by

(14) $$U(A \lor B) = \frac{U(A)P(A) + U(B)P(B)}{P(A) + P(B)} \text{ if } P(AB)$$
$$= 0 \neq P(A \lor B),$$

217

which is derivable from (9) and the fact that P is a finitely additive probability measure. Putting $\bar{A}$ for B in (14) we have

(15) $$U(T) = U(A)P(A) + U(\bar{A})P(\bar{A})$$

or setting $P(\bar{A}) = 1 - P(A)$

(16) $$P(A) = \frac{U(T) - U(\bar{A})}{U(A) - U(\bar{A})} \quad \text{if } U(A) \neq U(\bar{A}).$$

Equation (16) may seem to reveal a confusion of judgments of fact with judgments of value in the present theory. To see that this appearance is deceptive, notice that by (11), the utility point function, f, can be chosen quite independently of the probability measure P, but that the values of the utility set function, U, will be determined by P as well as by f, for the utility of a proposition is the probability-weighted average of the utilities of the ways in which it might come true. The relationship (16) is no more objectionable than the corresponding relationship in Ramsey's theory,

$$P(E) = \frac{U[C_1, E, C_2] - U(\bar{E}C_2)}{U(EC_1) - U(\bar{E}C)} \quad \text{if } U(EC_1) \neq U(\bar{E}C_2),$$

which expresses the probability of an event in terms of the utilities of a gamble on that event and of the two possible outcomes of the gamble.

In a sense, any proposition is a gamble, and in particular, given any proposition A which is preferred to its denial, the necessary proposition, $A \vee \bar{A}$, is a "natural" gamble on A in which the gain if the agent wins (if A is true) is the truth of A, and the loss if the agent loses (if A is false) is the truth of $\bar{A}$. It is only such natural gambles that are posited in the present theory. We do not assume, as Ramsey does, that given any pair of consequences C_1, C_2, and any event E, there is a gamble $[C_1, E, C_2]$ on E in which the agent stands to gain C_1 or lose C_2. Thus, consider the propositions that (C_1) there will be fine weather next week, (C_2) there will be a thermonuclear war next week, and (E) the next card dealt in the Principality of Monaco will be red. In Ramsey's system, you are supposed to be in a definite state of preference or indifference between the gamble,

fine weather next week if the next card dealt in the Principality of Monaco is red, thermonuclear war next week if not,

and (say) the consequence of breaking your ankle tomorrow while playing tennis. But to seriously entertain the possibility of this gam-

218

ble you would need to alter your notions of the causes of weather and war so radically as to deprive your state of preference or indifference between it and the consequence in question of all relevance to the world as you actually take it to be. In confining ourselves to "natural" gambles we work with the world as agents see it, including only those causal connections that they actually believe obtain among the propositions in their preference rankings.

It is of some interest to explain the notion of preference with which we are dealing in terms of an ideally self-knowing agent who is being offered options by a person whom I shall call "the operator." To determine the agent's state of preference as to the propositions A, B, the operator offers the agent his choice of (i) having A come true, or (ii) having B come true. If the agent chooses (i), B may come true, or it may not, but A will certainly come true. Similarly, if the agent chooses (ii), A may or may not come true, but B certainly will. The agent must choose (i) or (ii) – one, but not both. Choice of (i) shows that A is at least as high as B in the preference ranking, and choice of (ii) shows that B is at least as high as A. If B is the necessary proposition, choice of (ii) is in effect a request that the operator let what will be, be – that the operator not interfere in the course of events. And choice of (ii) rather than (i) then shows that A is no higher than the necessary proposition in the agent's preference ranking.

It is plausible to define A as *good, bad,* or *indifferent* accordingly as A is *above, below,* or *with* T in the agent's preference ranking. By (15) it is easy to see that if A is good while $\bar{A}$ is indifferent, then A must have probability 0, and that if two propositions are ranked together but not with T, then the more probable of them is the one whose denial is the further from T. Thus, if the preference ranking is

A, B
T
$\bar{A}$
$\bar{B}$

it must be that B is more probable than A. This will give some sense of how the probability ranking is deducible from the preference ranking. The deduction makes heavy use of the fact that the probabilities of propositions play a role in determining the preference ranking of their denials and of other compounds. Our notion of preference already involves the notion of probability; but of course it must, if the notion of probability is to be gotten from it.

To see the exact extent of this "circularity," notice that in explaining preference in terms of the agent's responses to choices offered by an operator, we must suppose that the operator is *neutral* in the sense of aiming neither to gratify nor to frustrate the agent. Suppose the preference ordering of possible states has a bottom rank. An operator whose object was to frustrate the agent as much as possible could be relied upon to make whatever happens happen in the worst possible way, and the agent's strategy in choosing between (i) and (ii) above would be to choose (i) if the worst state in A is better than the worst state in B, and to choose (ii) if the worst state in B is better than the worst state in A. The necessary proposition would then be at the bottom of the preference ranking, for it certainly contains the worst states. And since for any proposition A, either A or its denial contains a worst state, either A or its denial will always be ranked at the bottom, with T. Our previous considerations about probability and utility will then go by the board. Thus, by (14), if A and B are incompatible propositions which are not ranked together and which do not have probability 0, then $A \vee B$ will be ranked strictly between A and B; but with a malevolent operator, $A \vee B$ should always be ranked with the worse of A, B.

Then the sort of operator we have in mind when we explain preference in terms of the agent's responses to choices offered by an operator must be described by saying that when the agent chooses (i), say, the agent believes that the operator will make A happen in a way that is compatible with the agent's existing beliefs: If $A_1, \ldots,$ A_n are n incompatible propositions whose disjunction is A, then in choosing (i), the agent's probability for the operator to make A_i happen is $P(A_i \mid A)$, where P is the agent's conditional probability measure. This would vitiate the talk of an operator if it were intended as *the interpretation* of the notion of preference, rather than as a partial explanation. In fact, I do not regard the notion of preference as epistemologically prior to the notions of probability and utility. In many cases we or the agent may be fairly clear about the probabilities the agent ascribes to certain propositions without having much idea of their preference ranking, which we thereupon deduce indirectly, in part by using probability considerations. The notions of preference, probability, and utility are intimately related;

and the object of the present theory is to reveal their interconnections, not to "reduce" two of them to one of the others.

One of the principal uses of the theory of preference is to help the agent see what her state of preference or indifference between two propositions must be, given that her preferences between other related propositions are such-and-such, and that she wishes to be consistent in the Bayesian sense. Thus, consider the thesis that if A is preferred to B then $\bar{B}$ must be preferred to $\bar{A}$. This thesis is false, but its falsity is not obvious. To see it, consider an example in which A is the proposition that the agent dies tomorrow, B is the proposition that the agent is dishonored tomorrow, and the agent is in fact a Roman matron who, subscribing to the slogan "Death before dishonor" will straightway kill herself if dishonored. For consistency with what we may plausibly take her other preferences to be, she must also prefer $\bar{A}$ to $\bar{B}$: She must prefer a guarantee of living through tomorrow to a guarantee of not being dishonored all day tomorrow, because she takes the probability of living through tomorrow without honor to be zero. In particular, she believes that there are just three real possibilities as to the joint truth and falsity of A and B, which she ranks as follows:

$$\bar{A}\bar{B}$$
$$A\bar{B}$$
$$AB$$

Then $\bar{A}$ is practically the same proposition as $\bar{A}\bar{B}$ and is therefore at the top of this fragment of her preference ranking, while $\bar{B}$, the disjunction of the top and middle-ranking of the three propositions, must be ranked below the top and above the middle, since all three propositions have positive probability.

Let me now consider the third idiosyncracy of the present system – the fact that any fractional linear transformation $U_1 \to U_2$ as in (6) preserves the preference ranking. To focus attention on essentials, let us arbitrarily set $U_1(T) = U_2(T) = 0$ and $U_1(G) = U_2(G) = 1$, where G is some good proposition of which the denial is bad. Then if (as in Ramsey's theory) the only preference-preserving transformations are linear, all utility values are now determined. But by (6), (7), and (8) we have

(17)
$$U_2 = \frac{(c + 1)U_1}{cU_1 + 1}:$$

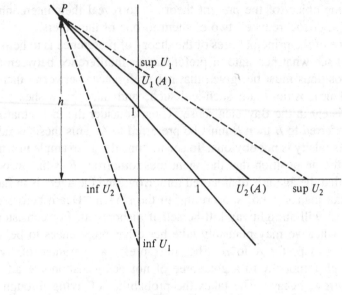

Figure 1.

There remains a certain freedom in choosing utility values, corresponding to the parameter c, which can be no smaller than -1.

The situation is shown graphically in Figure 1, where we suppose that c is negative. (The corresponding situation for positive values of c is that in which the point P is below inf U_1.)

In Figure 1, h, the height of the perspective point P, is $-1/c$, and the correspondence between the U_1 scale (vertical axis) and U_2 scale (horizontal axis) is simply this: that for each A, $U_2(A)$ is the perspective image of $U_1(A)$ from P. Since P lies on the line through the units of the two axes, we have $U_1(A) = 1$ if and only if $U_2(A) = 1$; and since the zeros of the two scales coincide we have $U_1(A) = 0$ if and only if $U_2(A) = 0$. But there are no other fixed points unless the perspective point is infinitely high, in which case the rays from P are parallel, at $-45°$, and we have $U_2(A) = U_1(A)$ for all A. This is the case in which h is infinite and therefore c is 0.

The startling fact is that by choosing h equal to the supremum of the U_1 scale we can transform a utility scale which has a finite upper bound into an equivalent utility scale that has no finite upper bound; and similarly (with P below the horizontal axis, on a level with inf U_1) we can transform a scale which has a finite lower bound into a

222

scale that has no finite lower bound. Boundedness above and boundedness below are not preferentially significant properties of utility functions in the present system. But we cannot transform a scale which is finitely bounded above or below or both into an equivalent scale which has neither a finite upper bound nor a finite lower bound; and therefore boundedness (above or below or both) is a preferentially significant property, even though the three separate species of boundedness are not.

This seems to put us into conflict with the renowned St. Petersburg paradox, as follows. Let $U(C_1)$, $U(C_2)$, . . . be an increasing, unbounded sequence, and for each n, let E_n be the proposition that the tail turns up first on the nth toss of a certain well-balanced coin. Then it is plausible that for each n, $P(E_n) = 2^{-n}$; and we can find a sequence m_1, m_2, . . . of positive integers such that for each i, $U(C_{m_i})$ is at least 2^i. The *St. Petersburg game* is an arrangement in which the agent tosses the coin in question until the tail turns up on, say, the tth toss, whereupon the consequence C_{m_t} is actualized. (Suppose that the consequences are payments of money to the agent.) Now the prospect of playing the St. Petersburg game has infinite utility; and the *St. Petersburg paradox* consists in the fact that in systems of the usual sort, a utility function which is unbounded above (below) must actually assume the value ∞ ($-\infty$). And the paradox is disturbing because it is unreasonable for the agent to be a Bayesian if there is a prospect of positive probability and infinite utility. Thus, let A, B, and C be propositions of positive probability; let AC and BC both have probability 0; and let the utilities of A, B, and C be u, v, w, where u and v are finite, w is infinite, and $u < v$. Then the agent, if reasonable, will prefer $B \bigvee C$ to $A \bigvee C$; but on Bayesian principles we have

$$U(B \bigvee C) = U(A \bigvee C) = w.$$

The usual way of solving the difficulty is to assume that the utility function is bounded above and below; but this move is not open to us, for boundedness is not a preferentially significant property of the utility function in the present theory. Instead, we deny that the prospect of playing the St. Petersburg game occurs in the agent's preference ranking, even if all the relevant E's and C's do. The basic remark is that anyone who offers to let the agent play the St. Petersburg game is a liar, pretending to have an infinite bank. We need

223

not (and cannot) assume that the utility function is bounded above and below; but we do assume that there are finite upper and lower bounds on the utilities of the prospects that it is within anyone's power to make happen at his pleasure. That is, for each person, we assume the existence of prospects E and F which are ranked above and below all prospects that are within that person's power to make happen at his pleasure. This is not open to Ramsey, who must assume that any two prospects can be the possible outcomes of a gamble.[8]

Finally, let us consider the fourth idiosyncracy: that probabilities are not completely determined by the preference ranking. For any proposition A there is a definite probability quantization, corresponding to the difference between the maximum and minimum values that $P_2(A)$ can be forced to assume by suitably adjusting c and d in (13). The difference between this maximum and minimum value for a given A is the preferentially significant quantity

$$(18) \qquad P(A)U(A) \left(\frac{1}{\sup U} - \frac{1}{\inf U} \right);$$

preferentially significant in the sense that it is invariant under all transformations (6), (13) of U and P. Furthermore, the least upper bound of this quantity, for all A in the preference ranking, will be

$$(19) \qquad (\sup I) \left(\frac{1}{\sup U} - \frac{1}{\inf U} \right)$$

Now whether or not U is bounded above or below, I will be bounded both above and below, and the coarseness of the quantization tends toward zero as $\sup U$ and $\inf U$ tend toward $+\infty$ and $-\infty$ with $U(T)$ fixed at 0 and $U(G)$ fixed at 1. Then in practice, the phenomenon of probability quantization need not be noticeable. Certainly it will make no difference to decision-making, since the quantization is precisely the degree of indeterminacy that is left over after all practical questions have been answered. The quantization will be zero if $\sup U = \infty$ and $\inf U = -\infty$, but I know of no reason to suppose that this must generally be the case for a rational agent, unless indeed it

---

8. In fact, the paradox need not arise in Ramsey's theory either, even if the utility function is unbounded, for no finite number of applications of the operation [,] will yield the St. Petersburg gamble. On the other hand, in Savage's system there are acts of infinite utility if the utility function is unbounded: see [8], p. 81.

be that only then are his probabilities uniquely determined by his preferences.

ACKNOWLEDGMENTS

This work was begun at Stanford University and completed at the Institute for Advanced Study and Princeton University. I wish to thank Rudolf Carnap and L. J. Savage for valuable criticism at the early stages. As noted above, the unicity problem was first solved by Ethan Bolker (in another context, independently of the present investigation); and he has since solved the mathematically deeper problem of characterizing the conditions under which a preference ranking is Bayesian. These results will appear in Bolker [1]. I am grateful to Bolker for his generosity in allowing me to read his unpublished work and for his patience and skill in explaining it to me. I am especially indebted to Professor Gödel for his generous help and encouragement during my term at the Institute. My first understanding of the mathematical situation described in the equivalence and uniqueness theorems came from him.

REFERENCES

1. Bolker, Ethan, *Functions resembling quotients of measures.* Ph. D. Dissertation, Harvard University, April 1965.
2. de Finetti, Bruno, *La prévision: ses lois logiques, ses sources subjectives. Annales de l'Institut Henri Poincaré,* vol. 7 (1937). English translation in [4].
3. Halmos, Paul R., *Lectures on Boolean algebras.* New York and London, 1963.
4. Kyburg, Henry E., Jr., and Howard E. Smokler, editors, *Studies in subjective probability.* New York and London, 1964.
5. Jeffrey, Richard C., *The logic of decision,* forthcoming [i.e., 1965]
6. von Neumann, John, and Oskar Morgenstern, *Theory of games and economic behavior.* Princeton, second edition, 1947.
7. Ramsey, Frank P., "Truth and probability," in *The foundations of mathematics and other logical essays.* London and New York, 1931. Reprinted in [4].
8. Savage, Leonard J., *The foundations of statistics.* New York and London, 1954.

225

14

Frameworks for preference

If I now prefer *p* to *q*, what sorts of entities are the prospects *p* and *q*? In 1954, Savage [10] answered this question in what he took to be behavioristic terms: Prospects – the terms of the preference relation – are functions from a set *S* of *states of nature* to a set *C* of *consequences*. He called such functions "acts." A decade or so later, Bolker and I answered it differently: Prospects are *propositions*, i.e., (nearly enough) sets of possible states of nature where the human agent is taken to be part of nature and his acts are thus ingredients in states of nature. Now, branching off from work of Luce and Krantz, Balch and Fishburn propose a hybrid answer: Prospects are act–event pairs and probability mixtures of such pairs. With Savage, they treat man and nature – acts and events – dualistically. Their treatment of acts is far more satisfactory than Savage's, and they are to be commended for the step toward holism which they take in dropping Savage's extraneous set *C* of consequences. But I will argue for the fully holistic or naturalistic position in which preference is a ranking of events (or propositions), some of which are acts.

With Balch and Fishburn, I find Savage's system unacceptable. Choice ought to reveal preference to at least this degree: The prospect which the agent chooses ought to be one of the highest in his preference ranking which he takes to be options for him. But in Savage's system, prospects seem to be entities of a sort among which finite beings cannot choose. For Savage, choice of an "act" is choice of a scheme which associates a definite consequence with each of the infinity of possible states of nature. If these are acts, then only God could know what act is being performed. After performance, the human agent may learn what consequence his act associated with the actual state of nature, but neither before nor after its

First published by R. Jeffrey, in *Essays on Economic Behavior under Uncertainty*, M. S. Balch, D. L. McFadden, and S. Y. Wu, eds. Copyright 1974 by Elsevier Science Publishers.

performance can he be expected to know what consequences the act associates with the rest of the possible states. Thus, for us, choice of an act cannot be choice of a particular function from states to consequences. To make Savage's scheme humanly applicable, one would have to modify it so as to make preference a relation between *sets* of functions from states to consequences. Choice of an action would then be choice of such a set, an unknown member of which will be realized. God *knows* which function is realized, and the human agent has *opinions* about the matter, expressed by a subjective probability distribution over the chosen set.

Balch and Fishburn seem to be making just this sort of complaint about Savage's theory in their footnote 3, where they contrast their theory with Savage's by temporarily formulating their proposal in something like his terms. The humanly available acts are various possible expenditures on advertizing, the consequences are various possible incomes from sales, and it is the *states* which are functions, namely, all functions from expenditures to incomes. Here human uncertainty has its proper object: not the identity of the act which is being performed, but the identity of the actual state of nature. I see this move as a definite conceptual improvement over Savage's representation of matters, and over the alternative just noted, of viewing preference as a relation between sets of Savage-style acts.

But Balch and Fishburn note this move only in passing, as a possibility. Their own move is to scrap Savage's consequences along with his functions from states to consequences. Hesitantly holistic, they take the basic prospects to be pairs (f, A) where f is a humanly possible act (not one of Savage's functions), and A is a subset of the set S_f of all possible f-conditioned states of nature (which need not be functions either). I take it that the agent prefers act f to act g when he prefers the pair (f, S_f) to the pair (g, S_g): Human choice is among unconditioned acts, i.e., among vacuously conditioned acts. The point of including such pairs as (f, A) and $(f, S_f - A)$ in the preference ranking is presumably to allow the agent to analyze his attitude toward f itself, i.e., toward (f, S_f), as a function of his degree of belief in f given A and of his attitudes toward f as they would be if he knew that the true state would be in A and if he knew that the true state would not be in A.

But why not be a bit more holistic, and view the agent as part of nature? A state of nature would then specify what act the agent performs, along with everything else one usually takes it to specify. Preference would be a relation on a Boolean algebra of subsets of the set S of all such holistic states. Underlying the preference ranking would be functions u and P on states and sets of states respectively: A would be preferred to B if and only if the conditional expected utility $E(u \mid A)$ of u on A were greater than $E(u \mid B)$, where both expectations are computed according to the probability measure P. Then prospects are propositions: sets of states. For the most part, prospects would be outside the agent's power to affect, for example, he might prefer fine weather tomorrow or rain tomorrow, even though there are no acts he can perform to realize either prospect. But among the prospects there would be certain propositions which he can make true or false as he pleases, and such propositions do duty as acts, for example, the act of taking his umbrella as he leaves the house in the morning would be represented by the set of states in which he does just that. Remember: The agent is part of nature, and his acts are ingredients in states of nature. In terms of the more conventional representation, in which the states of nature do not specify the agent's acts, my set S might be thought of as the cartesian product of the set of states with the set of acts (where the acts are not thought of as functions). But I prefer to think of acts as ingredients in states *ab initio*.

Such is the view which I put forth in my book and article of 1965 [6,5]. The thing would have been impossible but for the prior mathematical work of Bolker, set forth in full in [2], condensed in [3], and presented without technical details in [4]. Bolker's existence and uniqueness theorems are novel and important from a measurement-theoretical point of view, as are his methods, which look to von Neumann's work on continuous geometries rather than to Holder's theorem on ordered groups.

It was the work of von Neumann and Morgenstern [9] which persuaded economists and statisticians that cardinal utilities do, after all, make sense, and it was the work of Savage [10] which persuaded them that subjective probabilities make sense. In each case, the process of persuasion took some time, and in each case there was prior work (by Ramsey and de Finetti) which might have done the job of persuasion but did not: It was von Neumann–Morgenstern and Savage who finally got the ear of the public. By now, one no longer has to earn the right to deliberate in terms of subjective probabilities and utilities by first rebutting the ordinalists of the

228

1930s. Therefore I think it proper to characterize the notion of an ideally satisfactory preference ranking as one for which there exist a random variable·u and a probability measure P relative to which the conditional expected utilities $E(u \mid \cdot)$ mirror preferences: A is at least as high as B in the ranking if and only if $E(u \mid A) \geq E(u \mid B)$. From this characterization one can deduce that the preference relation is transitive, connected, etc. (Under "etc." we have the averaging condition: If A and B are disjoint prospects, then their union lies in the closed interval between them, in the preference ordering.) One might take the set of all such consequences to be the general theory of preference. This is not to deny the importance of existence theorems like Bolker's, which give conditions on the preference ranking and on the algebra of prospects from which one can deduce the existence of functions u and P as above. On the contrary: It is because we have such existence theorems that the foregoing procedure seems feasible. Note, however, that Bolker's conditions are not intended as axioms for the general theory of preference. Those conditions restrict the algebra of prospects in important ways, and make certain special assumptions about the preference ranking, so that their consequences neither exhaust nor lie wholly within the general theory of preference as defined above. But that is inevitable: One cannot expect the conditions to be necessary as well as sufficient for existence of the functions u, P.

But there remains a uniqueness problem, even if we sidestep the existence problem as I have suggested above. Consider the preference relation which is determined by a particular pair u, P, where u is bounded above or below. One might expect that any other pair u', P' which determined the same preference relation would be related to the first pair by the conditions $P' = P$ and $u' = au + b$ where a is positive. But in fact these strong conditions do not hold. In fact (Bolker's uniqueness theorem) the relevant group of transformations for u is not simply positive affine: It is a certain more comprehensive subgroup of the projective transformations (with positive determinant). Nor is P uniquely determined by the preference relation: There will be a certain "quantization" or uncertainty about the probabilities of propositions which appear above or below S in the preference ranking.

This underdetermination of u and P by the preference relation (unless u is unbounded above and below) is fascinating, but may be

229

seen as a flaw. (Must one have preferences of unlimited intensities in order to have a perfectly sharp subjective probability measure?) If so, the flaw is removable, for example, by using *two* primitives: preference and comparative probability. With these primitives, one ought to be able to drop some of Bolker's restrictions on the algebra of prospects in favor of conditions on comparative probability and conditions connecting preference and comparative probability. I would expect that in this way one could get significantly closer to an existence theorem in which the conditions are necessary as well as sufficient for existence of u and P, while obtaining the usual uniqueness result: P is unique, and u is determined up to a positive affine transformation. It would be a job worth doing.

To see the situation clearly, let us now think of prospects as probability measures over a measure algebra of subsets of S. On S there is a fixed random variable u which assigns utilities to possible states, and one prospect is preferred to another when the (unconditional) expected utility of u is greater when computed according to the first probability measure than when computed according to the other. If we take the set of prospects to include all probability measures over the underlying measure algebra, the preference relation determines u up to a positive affine transformation as in the von Neumann–Morgenstern theory. But in the Bolker–Jeffrey theory, the prospects form a much thinner set than that: They stand in one-to-one correspondence with the probability measures $P(\cdot | A)$ which are obtainable from a fixed measure P (the agent's actual subjective probability measure) by conditionalization relative to the various measurable subsets A of S to which P assigns positive values. [Note that the unconditional expectation of u relative to the measure $P(\cdot | A)$ is the conditional expectation of u on A relative to the measure P.] Preferences among prospects in this thin set determine u only up to a wider set of transformations, and the basic measure P is not uniquely recoverable from preferences if u is bounded above or below.

But to get the stronger determination of u, one need not fatten the set of prospects very much: It would be enough to have one additional prospect Q which is not of form $P(\cdot | A)$ for any measurable subset A of S but which is known to be (say) a 50–50 mixture of two such prospects: $Q = 1/2P(\cdot | A) + 1/2P(\cdot | B)$ for measurable subsets A and B of S where $P(A) \neq 0 \neq P(B)$ and $P(\cdot | A)$ is preferred to $P(\cdot | B)$. One would then know that the expected value of u relative to Q

230

is exactly half way between the expected values of u relative to $P(\cdot\mid A)$ and to $P(\cdot\mid B)$ and could use this fact to determine P uniquely and determine u up to a positive affine transformation. In these terms, the Balch–Fishburn sort of move would be to fatten the Bolker–Jeffrey set of probability measures by closure under *all* mixing operations: $aP(\cdot\mid A) + (1 - a)P(\cdot\mid B)$ with $0 \leq a \leq 1$ would be a prospect whenever $P(\cdot\mid A)$ and $P(\cdot\mid B)$ are.

These comments have been frankly tendentious and partisan. With Bolker, I applaud the Balch–Fishburn constructions as a real advance over Savage's approach, and over that of Luce and Krantz insofar as the latter continues to treat acts as functions, albeit partial functions, on a set of act-free states of nature. But I see the step from the Balch–Fishburn framework to Bolker's and mine as a further, natural advance to a truly holistic standpoint. If I have been noisy in my advocacy it is because that standpoint has not previously been called to the attention of foundationally minded economists and because Bolker's methods have not yet found their place in the toolkits of measurement theorists.

REFERENCES

1. Balch, M. and Fishburn, P. Subjective expected utility for conditional primitives. In Balch, McFadden, and Wu (1974), pp. 57–69.
2. Bolker, E. D. *Functions resembling quotients of measures.* Dissertation, Harvard University (April 1965).
3. Bolker, E. D. Functions resembling quotients of measures. *Transactions of the American Mathematical Society*, **124** (1966), 292–312.
4. Bolker, E. D. A simultaneous axiomatization of utility and subjective probability. *Philosophy of Science*, **34** (1967), 333–340.
5. Jeffrey, R. C. New foundations for Bayesian decision theory. In Y. Bar-Hillel (ed.). *Logic, Methodology and Philosophy of Science*. North-Holland: Amsterdam (1965), 289–300. [Essay 13]
6. Jeffrey, R. C. *The Logic of Decision*. McGraw-Hill: New York (1965).
7. Luce, R. D. and Krantz, D. H. Conditional expected utility. *Econometrica*, **39** (1971), 253–271.
8. von Neumann, J. *Continuous Geometry*. Princeton University Press: Princeton (1960).
9. von Neumann, J. and Morgenstern, O. *Theory of Games and Economic Behavior*. Princeton University Press: Princeton (1944).
10. Savage, L. J. *The Foundations of Statistics*. Wiley: New York (1954).

15

Axiomatizing the logic of decision

Ethan Bolker's theorem [1, 2] on the representation of functions resembling quotients of measures underlies the logic of decision [5] in the same way in which Hölder-like theorems on the representation of Archimedean ordered groups, semigroups, etc., are seen in [6] as underlying various foundational systems of measurement. In [3] and in chapter 9 of [1] Bolker applies his theorem to a system akin to that of [5], proving existence and uniqueness of probability and utility functions for preference rankings that satisfy certain axioms. Here I want to clarify the framework of [5] and bring Bolker's theorem to bear directly upon it. It should be noted that Bolker's theorem predated [5] and made it possible.

The logic of decision can be viewed as a theory with a binary relation term ("nonpreference") as its only primitive (apart from the notation of set theory or, alternatively, of higher-order logic). A *model* of the theory is a pair (u, P) where P is a probability measure on a σ- field M of "events" (measurable sets) and u is an integrable function on the union, W, of M. A sentence of the theory is valid in a model iff true for all assignments to its free variables of sets in M, when for $E \in M$ we define $U(E)$ as the conditional expected utility.

(1) $U(E) = \dfrac{1}{P(E)} \int_E u(w) \mathrm{d}P(w)$

   if $P(E) \neq 0$,

   $U(E) = U(W)$ if $P(E) = 0$,

and we define nonpreference in terms of U:

(2) $E \leqslant F$ iff $U(E) \leqslant U(F)$ for all $E, F \in M$.

Universal validity is validity in all models. But we shall not want

First published by R. Jeffrey, in *Foundations and Applications of Decision Theory*, Vol. 1, C. A. Hooker, J. J. Leach, and E. F. McClennen, eds., 1978. Reprinted by permission of Kluwer Academic Publishers.

to axiomatize universal validity, e.g., because models in which
$\leqslant$ holds between every two elements of M are of no decision-
theoretical interest. The aim is rather to axiomatize the notion of
validity in all models that are "nontrivial" in some suitable sense.

Call a relation $\leqslant$ of nonpreference *representable* when there exist
u and P for which (2) holds, with U defined as in (1). Such a pair is
then said to *represent* $\leqslant$. Define $I(E) = U(E)P(E)$ for E in M and
notice that I is a signed measure that vanishes wherever P does.
Then by the Radon–Nikodym theorem the pair (I, P) determines
functions u on W that satisfy (1), and any two such functions differ
only within a set of P-measure 0. Then (U, P) will also be said to
represent $\leqslant$ when (2) holds and $I = UP$ is a signed measure on M.

Notice that by additivity of I and P we have

(3) $\qquad U(E \cup F) = \dfrac{U(E)P(E) + U(F)P(F)}{P(E) + P(F)}$

$\qquad\qquad$ if $P(E \cup F) \neq 0 = P(E \cap F)$.

Define equivalence ($\approx$) as mutual nonpreference, and define strict
preference ($>$) as non-non-preference: $E \approx F$ iff $E \leqslant F \leqslant E$, and E
$> F$ iff $E \not\leqslant F$. Define *nullity*:

(4) $\qquad E \in N$ (E is "null") iff $E \cup F \approx F \not\approx E$ for some
$\qquad\qquad F$ disjoint from E.

For representable $\leqslant$, (A3) below ensures that the null events are
those of probability 0. This follows from (3), e.g., as in section 7.4
of [5]. Define *probivalence* (equiprobability of equivalent events):

(5) $E \doteq F$ (E and F are "probivalent") iff $E \approx F \not\approx G$ and $E \cup G \approx$
$F \cup G$ for some non-null G disjoint from E and from F.

By (3), $E \doteq F$ implies $P(E) = P(F)$ in case (U, P) represents $\leqslant$: See
[5], section 7.5.

Representation theorem (Bolker, 1965)

(Existence) A continous averaging function φ on a complete, atomless Boolean algebra S is isomorphic to a quotient $U = I/P$ of
measures (where I is a signed measure) iff φ is impartial ([1], 8.6;
[2], 6.12, 1.7).

233

(Uniqueness) Let f be an increasing function on the range Γ of U on $S - \{0\}$. Then $f \cdot U$ is a quotient of measures iff f is the restriction to Γ of a fractional linear transformation with positive determinant whose pole lies outside Γ. ([1], 5.10; [2], 3.6, p. 302 (printed out of order), where "with positive determinant" is omitted.)

Definitions:

(5) φ is an *averaging function* on S iff φ is not constant, has $S - \{0\}$ as its domain, and has its values in a linearly ordered set where $\varphi(A \vee B)$ lies in the interval from $\varphi(A)$ to $\varphi(B)$ if $A \wedge B = 0$, with endpoints excluded if $\varphi(A) \neq \varphi(B)$.

Here 0 is the zero element of S, and $\vee$ and $\wedge$ are the Boolean sum and product on S. In the definition of continuity,

A *chain* is a subset of $S - \{0\}$ that is linearly ordered by the Boolean order relation on S.

(6) φ is *continuous* iff whenever φ maps the supremum (or infimum) of a chain into an interval, it maps into that interval all members of the chain following (or preceding) some member.

[See [1], p. 102, V. If the range of φ has no extreme members relative to its linear ordering, the intervals in (6) may be taken to be open.]

(7) φ is *impartial* iff whenever $A \wedge B = A \wedge C = B \wedge C = 0$ and $\varphi(A) = \varphi(B) \neq \varphi(C)$ and $\varphi(A \vee C) = \varphi(B \vee C)$ we have $\varphi(A \vee D) = \varphi(B \vee D)$ for all D such that $D \wedge A = D \wedge B = 0$.

Axioms for the logic of decision. Think of $\leqslant$ as a set of ordered pairs (E, F). Then M below is the set of all members E, F, G, H, E_i of pairs in $\leqslant$, and W is the union of M.

(A1) M is a $\sigma-$ field, on which $\leqslant$ is transitive and connected.
(A2) $\emptyset \approx W$.
(A3) $G > W > W - G$ for some G.
(A4) If $E \doteq F$ and $E \cap H = F \cap H = \emptyset$ then $E \cup H \approx F \cup H$.

234

(A5) Every disjoint subset of $M - N$ is countable.

(A6) N is a proper $\sigma-$ ideal of M.

(A7) Every member of $M - N$ is the union of two disjoint members of $M - N$.

(A8) $E \approx F$ if $(E - F) \cup (F - E) \in N$.

(A9) For $E \cap F = \emptyset$: (a) $E \preccurlyeq E \cup F \preccurlyeq F$ if $E \preccurlyeq F$, and
(b) $E \not\approx E \cup F \not\approx F$ if $E \not\approx F$ and $E, F \in M - N$.

(A10) If $E_1, E_2, \ldots$ is a monotone ω-sequence with limit E and $F > E > H$ then $F > E_i > H$ for all i after some n.

Except for (A3) and (A7), all of these axioms are universally valid. Therefore, their redundancy is fairly harmless. But as axioms or theorems, (A1)–(A10) record properties of $\preccurlyeq$ that are useful in deriving the following representation theorem from Bolker's.

(8) Existence. $\preccurlyeq$ *is representable if (A1)–(A10) hold.*

(9) Uniqueness. *If (U, P) represents the $\preccurlyeq$ of (8) then (V, P) does iff there exist real a, b, c, d where $ad > bc$ and $-d/c$ is outside the range of U on M, and for all $E \in M$ we have*

$$V(E) = \frac{aU(E) + b}{cU(E) + d}, \; Q(E) = P(E)(cU(E) + d).$$

Existence proof. The quotient M/N of the σ-field M (A1) by its proper σ-ideal N (A6) is necessarily a complete Boolean algebra: See [4], p. 59. By (A7), no member of $M - N$ is an atom of M, and therefore M/N is atomless. Its members are subsets of M of form $[E] = \{F: (E - F) \cup (F - E) \in N\}$. In Bolker's theorem we set $S = M/N$, so that $0 = N = [\emptyset]$, $[E] \wedge [F] = [E \cap F]$, $[E] \vee [F] = [E \cup F]$, and the Boolean order relation $\leq$ on S is defined: $[E] \leq [F]$ iff $E - F \in N$. By (A8)(a), equivalence classes $\{F: F \approx E\}$ are unions of members of M/N, so that we can define a function v ("value") on M/N by writing

(10) $\quad\quad\quad\quad\quad v([E]) = \{F: F \approx E\}$

and we can define $>$ on the range of v:

(11) $\quad\quad\quad\quad\quad v([E]) > v([F])$ iff $E > F$.

By (A1), $\approx$ is an equivalence relation of M, and therefore the range of v is linearly ordered by the $>$ of (11). By (A3) and (A9), v is then an averaging function of M/N. By (A4), v is impartial. Finally, by (A10), v satisfies the definition (6) of continuity if all chains in M/N are countable, as they are by (A5), which says that M/N satisfies the "countable chain condition" ([4], p. 61): In a Boolean algebra satisfying the countable chain condition, every chain [in the sense of (6)] is countable. (See [4], p. 62, where Lemma 2 guarantees that every increasing chain [in the sense of (6)] is countable, and note that the elements A of any decreasing chain can be put into one-to-one correspondence with the elements A' of an increasing chain.)

Thus the v of (10), with range linearly ordered by the $>$ of (11), is a continuous, impartial averaging function on M/N, which is a complete, atomless Boolean algebra. Then by Bolker's existence theorem, v is isomorphic to a quotient $U = I/P$, in the sense that

(12) $v([E]) > v([F])$ iff $U([E]) > U([F])$ for all $[E]$,
$$[F] \in M/N.$$

Now transfer U to M by defining

(13) $U(E) = U([E])$ for all $E \in M$.

Applying (11) and (13) to (12), we have (2). Transferring P from M/N to M via

(14) $P(E) = U([E])$ for all $E \in M$

we finally establish the existence of a representation (U, P) of $\leqslant$ when (A1)–(A10) hold: The proof of (8) is complete.

Uniqueness of the representation as in (9) is then an immediate consequence of Bolker's uniqueness theorem for v on M/N.

In [3] (a version of chapter 9 of [1]), Bolker axiomatizes a closely related system, essentially as follows, where I indicate (loosely) the correspondence between his axioms and ours:

(B1) S is a complete (A6), atomless (A7) Boolean algebra. $\leqslant$ is transitive and connected (A1) on $S - \{0\}$ (A8), and continuous (A5, A10).

(B2) Averaging condition: (A9), written for $\leqslant$ on $S - \{0\}$.

(B3) Impartiality: (A4), written for $\leqslant$ on $S - \{0\}$.

Our definition (6) of continuity is from [1], p. 102; in [3], p. 337, Bolker defines it in the style of (A10) but allowing the monotone ωsequence there to be any monotone *sequence* (viz., *well-ordered sequence*).

I can identify some of the redundancies in my axioms, e.g., in place of (A6) it is enough to say that N is closed under countable unions and pairwise intersections, for by (A1)–(A3), ϕ is null but W is not.

REFERENCES

1. Bolker, E., *Functions Resembling Quotients of Measures*, Dissertation, Harvard University, April 1965.
2. Bolker, E., "Functions Resembling Quotients of Measures," *Trans. Amer. Math. Soc.* **124** (1966), 292–312.
3. Bolker, E., "A simultaneous axiomatization of utility and subjective probability," *Philosophy of Science* **34** (1967), 333–340.
4. Halmos, P. R., *Lectures on Boolean Algebras*, Van Nostrand, Princeton, New Jersey, 1963.
5. Jeffrey, R. C., *The Logic of Decision*, McGraw-Hill, New York, 1965.
6. Krantz, D. H., Luce, D., Suppes, P., and Tversky, A., *Foundations of Measurement*, Volume 1, Academic Press, New York, 1971.

16

A note on the kinematics of preference

In *The Logic of Decision* [4], preference is treated as a relation between propositions, viz., subsets of some set W of "all possible worlds." Such a preference ranking is called *coherent* iff representable by a pair u, P where u ("utility") is defined on W and P ("probability") is a probability measure on a σ-field of subsets of W relative to which u is measurable. To say that the pair u, P *represents* a preference ranking $\geqslant$ (where $>$ is strict preference and $\approx$ is indifference) is to say that preference goes by average utility, i.e., that we have

(1)
$$U(A) > U(B) \quad \text{if } A > B,$$
$$U(A) = U(B) \quad \text{if } A \approx B,$$

when u ("average" or "conditional expected" utility) is defined.

(2)
$$U(A) = \frac{1}{P(A)} \int_A u \, dP \quad \text{if } P(A) \neq 0.$$

Equivalent to (2) is the requirement that $I = UP$ be a signed measure. As an immediate consequence of (2) we have

(3)
$$U(A \cup B)P(A \cup B) = U(A)P(A) + U(B)P(B) \quad \text{if}$$
$$A \cap B = \varnothing.$$

Conditions (1) are framed so as to be compatible with the possibility that the domains of P and U are broader than the field of the relation $\geqslant$ of preference-or-indifference. Where a pair u, P represents $\geqslant$ we shall also say that the pair U, P does.

If a pair u, P represents $\geqslant$, so does any pair v, Q that is obtainable from it by the following sort of

---

First published by R. Jeffrey, in *Erkenntnis*, Vol. 11, pp. 135–141, 1977. Reprinted by permission of Kluwer Academic Publishers.

Rescaling.

$$v(w) = \frac{au(w) + b}{cu(w) + d}, \ Q(A) = P(A)(cU(A) + d).$$

$$V(A) = \frac{1}{Q(A)} \int_A v \, dQ = \frac{aU(A) + b}{cU(A) + d} \text{ if } Q(A) \neq 0,$$

where a, b, c, d are real constants with $ad - bc > 0$, $cU(W) + d = 1$, and $-d/c$ outside the range of u.

(This is part of the mathematics of the present theory that is taken over *in toto* from Ethan Bolker's [1, 2] abstract theory of generalized averaging functions. For an elementary exposition of the rescaling theorem, see [4], chapter 6.)

In example 5.3 of [4] notice is taken of an apparent difficulty in applying to decision theory the present theory of preference, in which acts may be represented by propositions, which have probabilities. (In general, acts and observations are to be represented by probability measures, but in the simplest cases these are obtained from the agent's subjective probability measure P by conditionalization on propositions that represent acts. For a discussion of somewhat less simple cases, see chapter 11 of [4].) But the discussion of this difficulty in [4] was inadequate, and recently, Wolfgang Spohn [5] and Ethan Bolker (in a letter) have taxed me with it. Let me now try to resolve it.

Perhaps the problem is seen most clearly when put in the form of a paradox, as follows:

Suppose you think it within your power to make A true if you wish, that you prefer A to W, and that you are convinced that A is preferable to every other one of your options. Then $P(A) = 1$, for you know you will make A true. But then, setting $B = -A$ in (3), we have $U(W) = U(A)$, so that by (1) you are indifferent between A and W after all, in contradiction to the assumption that you prefer A to W.

In a nutshell, in prose: In the light of your awareness of your options, your awareness of your preference for your top-ranked option over W reduces that preference to indifference. (Here W need not be an option, but it may be. To opt for W is to choose not to intervene in the course of events: To let what will be, be.)

239

To resolve this paradox, note that our preferences do and should change in various ways, from time to time, and that these modes of change can be described in terms of two polar types: Valuational (change in the utility function, u) and doxastic (change in the probability function, P). Here I focus on the second type: The underlying utility function u is held fixed, but beliefs change, and so P changes, and, with it, U.

In fact, I shall consider only the simplest sort of doxastic change: That in which preferences change simply because belief in some proposition A has changed from a nonextreme value $P(A)$ to 1. Before the change, values and beliefs were represented by functions u, P. After the change, they will be represented by functions u, P_A, where the utility function is unchanged but the probability function has changed from P to the conditional probability, $P_A(B) = P(A \cap B)/P(A)$. Then the average utility function that is determined by u and P_A as in (2) will be given by

$$(4) \qquad U_A(B) = \frac{1}{P_A(B)} \int_B u \, dP_A = U(A \cap B).$$

The change in subjective probability from P to P_A may arise because one has observed that A is true or for other reasons, notably because one has *decided that A shall be true*. (Then the act is *making A true*, and A need not be the proposition that such-and-such an act is performed.)

Now to resolve the paradox, observe that the alleged contradiction arises through conflation of two different preference rankings: yours before deciding to make A true, and yours thereafter. Before the decision, your preferences were $\geqslant$, represented by U, P. Afterward they were $\geqslant_A$, represented by U_A, P_A. The argument purporting to prove that after all, $U(A) = U(W)$, actually proves nothing but the triviality $U_A(A) = U_A(W)$, i.e., by (4), $U(A) = U(A)$.

Deliberation – deciding what to do – is a matter not only of clarifying your preferences, but also of identifying your options. Clarity about your preference for A over W (or over $-A$ or over any other proposition) does not mean that $P(A) = 1$, for you may not (yet) see A as in your power to make true if you choose, and you may not (yet) see A as the highest proposition of that character in your preference ranking. And even if you are convinced that (say) your options are A, W, and $-A$, and that A is fully in your power to make true, it may be that $P(A) \neq 1$ because you consider that before the time comes to enact A, your preferences may change for some reason or other so that when the time comes, A may no longer be the highest-ranked of your three options. For all of these reasons, present recognition that A is your top-ranking option and that you can make A true if you wish does not entail $P(A) = 1$, viz., present full belief in A.

But of course if matters stand so when the time for action comes, and if your preferences *at that time* are given by P, U, *then* $P(A) = 1$ and $U(A) = U(W)$ and A is *not* then preferred to W. But that is at the time for action, when $P(A) = 1$ because you know you are about to enact A – a thing you are about to do because of your *preference* for A until that crucial moment. Could your indifference between A and W as you enact A lead you to abort that action, and let what will be, be? Of course not: That indifference is predicated on the very fact that you *are* enacting A (and in a way that you are sure cannot fail).

I am indebted to David Lewis for identifying the source of the difficulty in my first attempt to resolve the paradox: I had been insisting both that

(5) The impossible proposition, $\varnothing$, occur at the same position as W in every preference ranking,

and that

(6) $\qquad U_A(B) = U(A \cap B)$ if $P(A) \neq 0$

for *all* B in the preference ranking, and not only when $P(A \cap B) \neq 0$, as in (4). But taken together, (5) and (6) are incompatible with serious preference: *Where (5) and (6) hold, the agent is indifferent among all propositions in which he has positive degrees of belief.* Proof. By (6), $U_A(\varnothing) = U(\varnothing)$ and $U_A(W) = U(A)$, and by (5), $U_A(\varnothing) = U_A(W)$ – if $P(A) \neq 0$. Then as long as $P(A)$ is positive, $A \approx \varnothing$, and so all such A are ranked together.

I had espoused (5) for a trivial reason: I wanted to call A "good," "bad," or "indifferent" depending on whether $A > -A$, $-A > A$, or $A \approx -A$, and by adopting (5) I ensured that W would count as indifferent. At the other extreme, Ethan Bolker [1, 2, 3] excludes from the preference ranking all propositions of probability 0. My present inclination is to include $\varnothing$ in every preference ranking, but cut it loose from W: Drop (5) but keep (6). In the sense that this move has no effect whatever on the outcome of any deliberation, it is pure fancy. But it is a fancy that fits well with the motivation behind (4), and that may help dispel any lingering air of paradox concerning acts of making propositions true.

A different fancy leads to choice of

(7) $V_A(B) = U(A \cap B) - U(A)$ if $P(A) \neq 0$

instead of U_A as one's new average utility function, when one's new belief function is P_A. Here we have $V_A(A) = V_A(W) = 0$, but we have $V_A(\emptyset) = V_A(W)$ only in the special case where $V_A(\emptyset) = 0$, i.e, by (7), where $U(A) = U(\emptyset)$.

The difference between U_A as in (6) and V_A as in (7) is mere rescaling, of the sort that has no effect on the preference ranking. If we suppose that whether we represent the new preference ranking by U_A or by V_A, the new belief function is P_A, then two of the scale constants are determined: $c = 0$, $d = 1$. The other two constants are determined as $a = 1$, $b = -U(A)$ by the fact that $V_A(B) = U_A(B) - U(A)$. It is easily verified that the three conditions on the scaling constants are then met.

Since the difference between V_A and U_A is only a matter of rescaling, the observation that $V_A(A) = V_A(W)$ implies that $U_A(A) = U_A(W)$, for in either case we have $A \approx_A W$ where $\approx_A$ is indifference, in the new ranking. Similarly we have $\emptyset \approx_A W$ iff $\emptyset \approx A$ in consequence of the observation about V_A made at the end of the paragraph containing (7), above. Then (5) does not become tolerable through rescaling. On the contrary, to insist upon (5) is to restore the paradox: According to (5), indifference between A and W in the new ranking implies that even in the old ranking there was indifference between A and $\emptyset$ and, by (5), W.

The paradox is *essentially* kinematical or diachronic. There is no harm in insisting that at a particular time, the agent be counted indifferent between $\emptyset$ and W. But if we insist that he be counted indifferent between $\emptyset$ and W at *all* times, we rule out all conditionalization except for that in which the proposition A that is conditionalized upon (in going from P to P_A) was ranked with W *before* the change. Thus I drop (5).

An alternative would be: Restrict (6) and (7) to propositions B for which $P_A(B)$ is positive. But this accords ill with the fancy (2) that U

242

(A) is the average of the values $u(w)$ as w ranges over A. Thus, consider the set $\{w: u(w) = U(A)\}$ for some A. This is the set of all those possible worlds that have as their utilities exactly the average of the utilities of worlds in which A is true. That set is the proposition that the utility of the real world is the average of the utilities of the worlds in which A is true. Such a proposition is unlikely to have positive probability, and yet it is clear what the average of $u(w)$ is, as w ranges over the worlds in which it is true. As all those worlds have the same utility, $u(w) = U(A)$, the number $U(A)$ must be that average. So *some* propositions of probability 0 have clear places in the preference ranking, if we take seriously the fancy underlying (2). (Of course, we needn't do so, e.g., it plays no role in decision-making. And with Bolker [1, 2, 3] one can even drop possible worlds from the framework by taking propositions to be elements of a Boolean algebra not representable as a field of sets.)

Why choose U_A as in (6) instead of V_A as in (7), to represent the preference ranking of one whose utility and probability functions have been u and P, but who then learns or decides that A is true? The reason is given by the calculation (4). If the change was purely doxastic, so that the utility function is the same after the change as it was before, its average after the change will be U_A, not V_A. V_A is the average of a different utility function, $v(w) = u(w) - U(A)$, $c = 0$, $d = 1$. To change from u to v as a result of determining that A is true is to change one's valuations of all possible worlds as a result of determining that A is true in the real world.

A final word about U_A. There is no need to have $\varnothing$ in the preference ranking, even if one is determined to take u seriously as in (2). But one might want to have $\varnothing$ in the ranking as a marker of where W once was, e.g., at a time when things looked neither particularly good nor particularly bad, but middling. If $\approx$ is one's indifference relation as it was at that time, one can set $\varnothing \approx W$ (no harm in making that stipulation *once*), and forever after, W will be ranked above, with, or below $\varnothing$ depending on whether at that later time one thinks that by and large one is better off, just as well off, or worse off than one was when one set $\varnothing \approx W$. That is, $\varnothing$ will serve as such a marker as long as one's preference ranking is representable as a result of conditionalizing, relative to some proposition A, the belief function one had when one set $\varnothing \approx W$. (Reason: $U_A(W) = U(A)$, U_A ($\varnothing$) $= U(\varnothing) = U(W)$. Then the position of W relative to $\varnothing$ in the new ranking will always be the same as that of A relative to W in the old ranking.) When one becomes indifferent between A and W as a result of taking steps to make A true, where A was preferred to W, the new indifference comes about not through a drop in the average utility of A, but rather through an increase in the average utility of W, for $U_A(W) = U(A) > U(W)$.

REFERENCES

1. Bolker, E. *Functions Resembling Quotients of Measures*, Dissertation, Harvard University, April 1965.
2. Bolker, E., "Functions Resembling Quotients of Measures," *Trans. Amer. Math. Soc.* **124** (1966) 292–312.
3. Bolker, E., "A Simultaneous Axiomatization of Utility and Subjective Probability," *Philosophy of Science* **34** (1967) 333–340.
4. Jeffrey, R., *The Logic of Decision*, McGraw-Hill, New York, 1965.
5. Spohn, W.: *Erkenntnis* **11** (1977), p. 113.

244